WOODWORKER'S GUIDE TO WOOD

RICK PETERS

•SOFTWOODS •HARDWOODS •PLYWOODS •COMPOSITES •VENEERS

Sterling Publishing Co., Inc.
New York

Butterick Media Production Staff

Design: Sandy Freeman
Cover Design: Elizabeth Berry
Photography: Christopher J. Vendetta
Cover Photo: Brian Kraus, Butterick Studios
Illustrations: Greg Kopfer, Triad Design;
   Elizabeth Berry
Technical Support: Jim Kingsepp

Copy Editor: Barbara McIntosh Webb
Page Layout: Sandy Freeman
Index: Nan Badgett
Assoc. Managing Editors: Stephanie Marracco,
   Nicole Pressly
Project Director: Caroline Politi
President: Art Joinnides

Every effort has been made to ensure that all the information in this book is accurate. However, due to differing conditions, tools, and individual skill, the publisher cannot be responsible for any injuries, losses, or other damages which may result from the use of information in this book.

Library of Congress Cataloging-in-Publication Data

Peters, Rick
   Woodworker's guide to wood: softwoods, hardwoods, plywoods, composites, veneers / Rick Peters
     p. cm.
ISBN 0-8069-3687-8
   1. Wood. 2. Lumber. I. Title.

TA419.P385 2000
684'.08—dc21   99–086641

Published by Sterling Publishing Company, Inc.
387 Park Avenue South, New York, N.Y. 10016

©2000, Butterick Company, Inc., Rick Peters

Distributed in Canada by Sterling Publishing,
c/o Canadian Manda Group, One Atlantic
Avenue, Suite 105, Toronto, Ontario, Canada
M6K 3E7
Distributed in Great Britain and Europe by
Cassell PLC, Wellington House, 125 Strand,
London WC2R 0BB, England
Distributed in Australia by Capricorn Link
(Australia) Pty. Ltd., P.O. Box 6651, Baulkham
Hills, Business Centre, NSW 2153, Australia

Printed in China
All rights reserved

Sterling ISBN 0-8069-3687-8

THE BUTTERICK® PUBLISHING COMPANY
161 Avenue of the Americas
New York, New York 10013

CONTENTS

ACKNOWLEDGMENTS

For all their help, advice, and support, I offer special thanks to:

Christopher Vendetta for scrambling over sawmills, wandering through forests, and coping with mountains of sawdust to take the photographs for this book.

Curt Alt of the Hardwood Plywood Veneer Association for providing photos of the hardwood plywood manufacturing process.

Dean Brandt and the expert wood recycling crew at Sylvan Brandt for letting us photograph their reclamation operation.

Earl Deemer, president of the IWCS (International Wood Collectors Society) for letting me tap into his network of wood lovers and experts.

Dick and Mary Deihl, talented, generous, and dedicated IWCS members who lent me their precious turned wooden eggs for the cover.

Jim Flynn, IWCS member, respected author, and wood guru, for proofreading the Nature of Wood and Directory of Wood chapters, and gently pointing out mistakes (particularly with the botanical names).

Erik Granberg of Granberg International, for assistance with a chainsaw mill and with technical information.

Gary Green, IWCS member and wood sample aficionado, for his willingness to drop what he was doing and ship me yet one more wood sample.

Carl Hunsberger and the crew at Hunsberger's Sawmill in Quakertown, Pennsylvania, for letting us photograph his mill in action and sharing his years of wisdom.

Mike and Louise Peters at Shady Hill Farm, for their willingness to fit me into their busy schedule and to cut any log any way I wanted, and for their suggestions on the Milling and Drying Lumber chapters.

Bill Powell from States Industries, for supplying product and technical information on ApplePly and prefinished plywood.

Reginald Sharpe of the Structural Board Association, for providing the illustrations on how oriented-strand board is made.

Beth Tyler with the Composite Panel Association, for supplying the illustrations that show how particleboard is made.

The folks at Butterick for their continuing support: Art Joinnides, Caroline Politi, Stephanie Marracco, and David Joinnides. Also Barb Webb, copyediting whiz, and especially Sandy Freeman, whose exquisite art talents are evident in every page of this book.

Heartfelt thanks to my family: Cheryl, Lynne, Will, and Beth, for putting up with the craziness that goes with writing a book and living with a woodworker: late nights, short weekends, wood everywhere, shop noise, and sawdust in the house. And finally, words can't express my gratitude to my wife Cheryl for taking off the rough edges of the manuscript.

INTRODUCTION

This is the book I needed 25 years ago, when I got started in woodworking.

I loved wood and wanted to work with it, but I got so caught up in trying to outfit a shop and learn a zillion woodworking techniques that I lost sight of the material. For the longest time, not knowing any better, I treated wood as if it were plastic. And the wood rebuked me: Glue joints failed, joinery popped apart, tabletops split. That's when I discovered that technique without wood knowledge can take you only so far. To work wood, you have to understand it.

So, I began learning. I learned about wood and wood products—mostly by trial and error— reasoning that the more I knew, the better my projects would turn out. And I was right. Now, my glue joints and joinery are rock solid, my tabletops seamless. There are no nasty surprises. And when a piece is done, I get to hear the phrase that warms a wood-

worker's heart: "You made that? Wow!"

There's no substitute for experience when it comes to working wood. Still, you don't have to burn your hand to see whether the oven is hot. My hope is that from these pages, you'll take the benefit of my experience and that of a host of experts, and apply it to your own learning. I also hope you'll learn faster and with fewer hard lessons than I did.

It's not until you've seen light reflecting off a beautifully quartersawn white oak panel that you can really appreciate ray fleck. Or until you've had a board pinch a saw blade that you can understand case-hardening. You can't savor the wonderful sweet smell of white ash from a book.

But it's always possible to become a better woodworker. I hope this book helps.

Rick Peters
Spring 2000

"Murmuring out of its myriad leaves,
Down from its lofty top rising over a hundred feet high,
Out of its stalwart trunk and limbs, out of its foot-thick bark,
That chant of the seasons and time, chant not of the past only
    but of the future."

WALT WHITMAN (1874)

The NATURE of WOOD

Whitman knew something of the spirit of wood. Like the tree he so admired, the wood it contained was organic, fluid, and elemental. And like the tree, wood does speak of the future—a unique future that requires respect. We work with it because we like its texture and feel...we savor the smell that rises from a cut...we're gratified by how the grain "pops" when a finish coat is applied.

We appreciate wood, and we admire the poetry-inspiring trees that produce it. But one thing about its organic nature can delight us—or make us wish we had taken up a simpler pastime. Unlike glass, ceramic, or steel— wood moves. Long after you've crafted it into furniture or fenceposts, wood continues to move as the moisture content of air changes. When you understand and allow for this in your woodworking, your pieces are strong and lasting. Nothing buckles or falls apart, because you know that wood moves.

To work wood successfully, you need to know why it behaves the way it does. You need to know a slice of what botanists know. Only then can the material you work so carefully pay back your attention with enduring quality. It's the nature of wood.

SOFTWOOD

All of the many beautiful woods available to woodworkers come from trees classified as either softwoods or hardwoods. Inexplicably, not all softwoods are soft and not all hardwoods are hard. For example, balsa, a hardwood, produces one of the softest woods in the world. Conversely, Douglas fir, a softwood, is harder and has better strength properties than many hardwoods. A clearer classification would be to describe softwoods as needle-bearing trees and hardwoods as leaf-bearing.

Conifers

There are four families of Coniferales within the softwood group. The term conifer is used to describe a tree that's characterized by needle-like or scale-like foliage—usually evergreen. Most softwood trees are identifiable by their cone shape, a dominant stem, and lateral side branching (*top photo*). Most of the world's commercial softwoods grow in the northern hemisphere.

Needles and naked seeds

Botanically, softwoods are gymnosperms, which in layman's terms means the seeds are naked—not enclosed in a flower (*inset*). You'll often see the seeds borne on the scales of a cone, resembling small berries like those of a juniper, or in a cup such as on a yew. Although most conifers are evergreens, Mother Nature threw in a twist: Some conifers, like the larch, drop their needles in the fall.

Appearance

When converted into lumber, most softwoods are relatively light in color and range from pale yellow to reddish brown. The annual growth rings formed by layers of earlywood and latewood (*see page* 15) are typically very easy to distinguish. Most softwoods are fast-growing and are harvested primarily for manufacturing and construction-grade lumber.

Hardwood

Hardwoods are generally heavier and denser than softwoods and so are much more sought-after by woodworkers for their projects. In addition, hardwoods offer a wide variety of color and grain options for the woodworker that softwoods just can't match (*see below*).

Deciduous

There are over 20 families of hardwoods in the United States alone. Most hardwood trees are deciduous, meaning their leaves fall off every autumn; this doesn't apply to all hardwoods, though. Those in tropical regions often keep their leaves year-round, and are evergreen. Most hardwood trees have a round or oval crown of leaves and a trunk that divides and subdivides (*top photo*). For the most part, broad-leaved deciduous hardwoods grow in the temperate northern hemisphere, and broad-leaved evergreens grow in the southern hemisphere and tropical regions.

Leaves and nuts

Botanically, hardwoods are angiosperms—the seeds are enclosed or protected in the ovary of the flower, typically a fruit or a nut. Protection can vary greatly, from the delicate skin of a pear to the tough shell of an acorn (*inset*).

Appearance

When converted into lumber, hardwoods offer a dizzying array of colors, figure, and grain. Colors range from the near white of holly to the jet black of ebony. Vibrant colors abound, such as the blood red of padauk, the bright orange of osage orange, and the deep violet of purpleheart. A combination of rays and vessels in hardwoods can produce wild effects in grain: ray fleck, tiger stripe, and fiddleback, to name a few (*see pages 98–99*). Growth rings of hardwoods can be difficult to distinguish; tropical hardwoods have virtually no growth rings since the growing season is year-round.

HOW A TREE GROWS

■ Seeds are dispersed from a mature tree in a number of ways: Some seeds have broad-bladed wings to catch the wind, others have hooks to grab insects or are scented to attract birds. Most seeds never make it to a fertile spot where the conditions are right for sprouting. If the soil is warm, rich, and moist, a root tip and shoot tip will soon emerge. The root tip burrows into the ground and, via tiny hairs, absorbs moisture and elements essential for life. At the same time, the shoot tip grows upward toward light so that photosynthesis can occur.

In its first year, the seedling will develop a special layer of regenerative cells called the vascular cambium, or simply the cambium. This important layer is only one cell thick and is what allows a tree to grow (*see page* 11 *for more on this*).

The cambium layer forms a complete sheath around the entire tree. As growth continues, new layers are added to the pith, which was formed by the original shoot. Side shoots, which ultimately become branches and finally knots, also begin in the pith. Sheathing the entire tree like this means that both height and girth are gained at the same time as annual increments of wood are laid down in a conelike form (*see the drawing at right*).

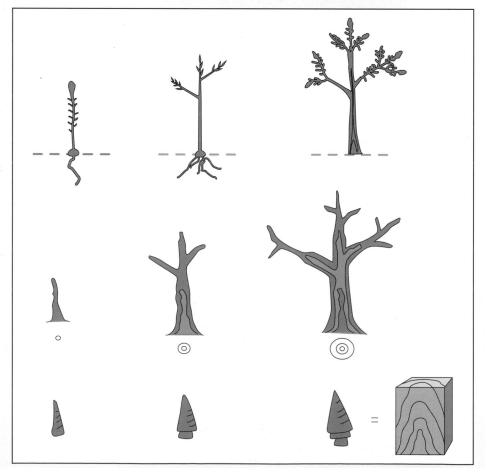

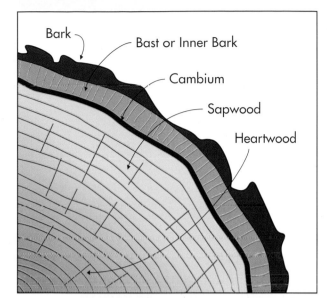

Bark
Bast or Inner Bark
Cambium
Sapwood
Heartwood

A look at the layers of a maturing tree (*top drawing*) shows what goes on inside. The outermost layer is bark, composed of an outer, corky, dead covering, and an inner, living bark or "bast." The inner bark carries food from the leaves to the growing parts of the tree. Sapwood transports sap from the roots to the leaves; heartwood is formed by a gradual change in the sapwood and is inactive (*see below*). Rays are horizontally oriented tissues that connect various layers from pith to bark for storage and transfer of food.

Cambium

Between the inner bark and the sapwood is the cambium, which actually forms the wood and bark cells. Cells divide frequently and each may become either a bark cell or a wood cell. Wood cells form on the inside of the cambium; bark cells form on the outside. New wood is laid down on top of old wood, and the diameter of the trunk increases. The existing bark is pushed outward by the new bark, creating its familiar cracked and stretched appearance.

Heartwood and sapwood

As the end of the growing season nears, the tree's large crown of leaves often produces more food than it can use (*middle photo*). This excess food (called photosynthate) moves from the inner bark through the rays to the center of the tree. Here it accumulates and over time breaks down to form compounds known as extractives, which plug up the cells and eventually kill them. This area of extractive-impregnated dead cells is the heartwood. Frequently, the extractives darken the heartwood and give it its characteristic color (*bottom photo*).

A cross section cut from a tree (*top photo*) shows a distinct boundary between heartwood and sapwood. The cross section shown here is of yellow poplar; the heartwood is easily recognizable by the green center section in contrast to the lighter sapwood.

Peeling the tree also helps to display the different layers and clearly define their proportions (*middle drawing*). Note: The only thing here not to scale is the cambium; remember, it's only a cell or two in thickness. Removing the dead, rough outer bark reveals the smooth, lighter inner bark, where food in the form of photosynthate is transported from the leaves to the center of the tree. Rays in the inner bark align with the rays in the sapwood and heartwood to funnel the fluids inward.

Growth rings

New wood cells that form early in the growing season are large and have thin walls. Later in the year, new cells are smaller and have thicker walls (*see page 14 for more on cell structure*). As the cells build on top of each other, layers form. Layers of early cells and late cells can be easily distinguished from each other by their width and color. Typically, the early cell layer is light and wide, and the layer of late cells is thin and dark.

In temperate regions, one layer of early cells and one layer of late cells define one growing season. This combination is referred to as an annual growth ring. The age of a tree can be quickly determined by counting the rings; the tree shown in the bottom drawing is seven years old. This tree-dating process is known in the science world as dendrochronology.

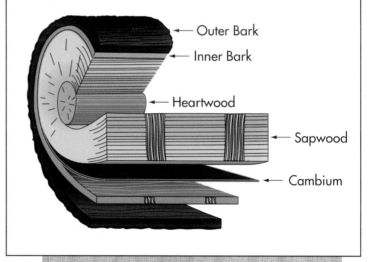

Outer Bark
Inner Bark
Heartwood
Sapwood
Cambium

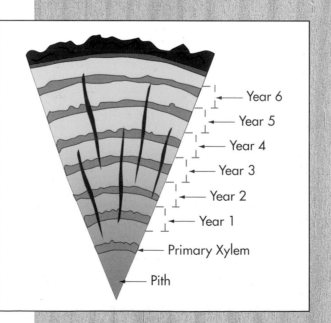

Year 6
Year 5
Year 4
Year 3
Year 2
Year 1
Primary Xylem
Pith

GRAIN

A woodworker can't talk about wood without the word *grain* popping up immediately. The problem is, grain means different things depending on how it's used. For example, "that wood has uneven grain," and "oak has coarse grain," and "use a block plane on end grain" all describe different characteristics of grain.

Grain can describe planes and surfaces, such as radial, tangential, and traverse (*top drawing*). It can describe growth-ring placement, such as edge grain, rift grain, and side grain, or growth-ring width (open grain and close grain), and even the contrast between earlywood and latewood (even grain and uneven grain). Grain can also illustrate the alignment of cells (across-the-grain, along-the-grain, with-the-grain), note the pore size (fine grain, coarse grain), express different types of figure (curly grain, roey grain, quilted grain), and define machining defects like chipped grain, fuzzy grain, and raised grain.

Technically, grain is defined as the direction of the wood fibers in a tree. A radial surface, and therefore radial grain, is created when you cut along the radius of a round cross section. When a log is cut like this, it is said to be quartersawn, and the grain is usually straight and uniform. Tangential grain is created by cutting at a tangent to the growth rings. When a sawyer cuts logs like this, it is called plain- or flat-sawing. A traverse or cross section is what you get when you cut perpendicular to the wood fibers. Crosscutting a log this way produces end grain.

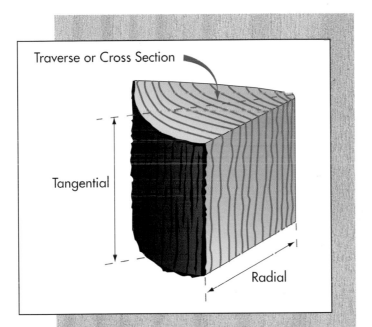

Traverse or Cross Section

Tangential

Radial

APPEARANCE

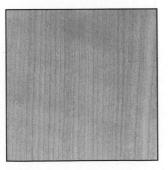

Radial: A radial-cut log will produce quarter grain, which is typically straight and uniform in width.

Traverse: A traverse-cut log will show end grain. Note the visible annual growth rings.

Tangential: When a log is cut tangentially, it produces plain-sawn or flat-sawn boards.

SOFTWOOD CELL STRUCTURE

There's one thing that distinguishes wood from all other crafts materials: cell structure. Because it's organic, wood continues to move with changes in humidity even after it has been cut and dried—it's hygroscopic (it absorbs and releases moisture). A little tree science will help you understand why it behaves the way it does, but you don't have to be a botanist to grasp cell structure. This is especially true with softwoods, whose structure is simple and uniform.

Cells

There are only two types of cells in softwood: wood fibers (technically, longitudinal tracheids) and ray cells (*top drawing*). About 95% of the softwood is made up of these wood fibers, which resemble soda straws. They're long, hollow, and pitted; but unlike a soda straw, they're tapered shut at both ends. The diameter of the fibers determines the wood's texture, surface smoothness, and finishing qualities. Large-diameter fibers result in coarse-textured wood; small-diameter fibers create a fine texture. Short strips of radially aligned ray cells make up the remaining 5% or so of the volume.

Resin canals

Resin canals are found in only a few softwoods, like pine, spruce, larch, and Douglas fir. They serve as a defense mechanism in the tree by transporting resin to an injury. Resin canals are spaces between softwood cells caused by separation of adjacent cells. When cells separate, special resin-producing cells called epithelial cells are formed (*red cells in middle drawing*).

Pits

Earlier I said wood fibers are tapered shut on the ends. If that's so, how does fluid move through the fibers? It travels through missing sections of the cell walls called pits (*bottom drawing*). Normally, the pit of one cell aligns with the pit of another cell to form a pathway (*inset*).

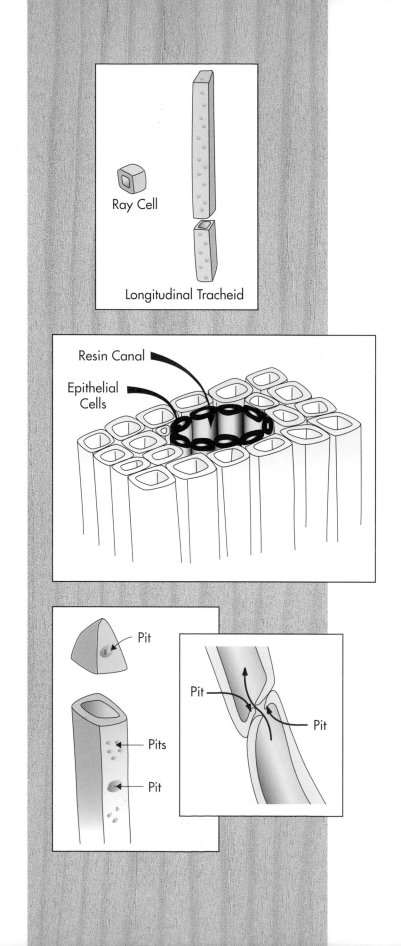

Ray Cell

Longitudinal Tracheid

Resin Canal

Epithelial Cells

Pit

Pits

Pit

Pit

Pit

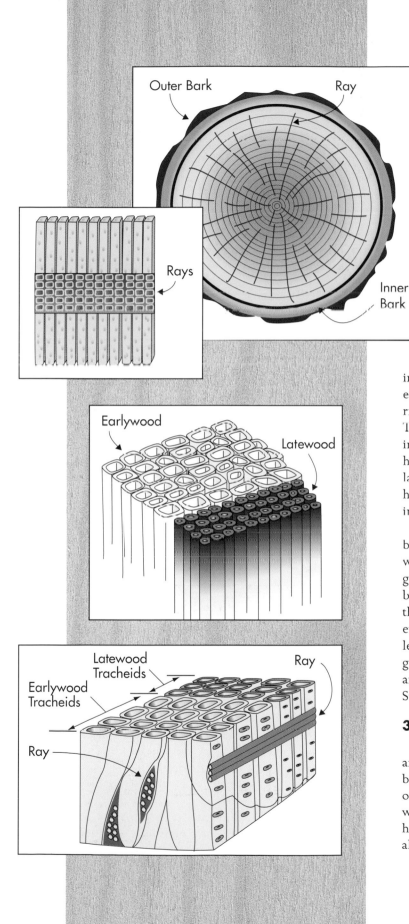

Rays

Wood fibers are arranged more or less longitudinally—that is, parallel to the trunk of the tree. Ray cells, on the other hand, are oriented radially or perpendicular to the trunk (*top drawing*). Ray cells are most commonly found in ribbons simply called rays (*inset*). Rays are basically roadways for transporting food and other materials around inside a living tree.

Earlywood and latewood

Regardless of where a tree is growing, it will always alternate growing periods with rest. In any region with a temperate climate, there is one growing period (spring and summer), followed by one rest period (fall and winter). This results in adding just one layer of new wood to the tree each year, commonly called an annual growth ring, an annual ring, or simply a growth ring. The sap-conducting wood cells that form early in the growing season (earlywood) are large and have thin walls (*middle drawing*). Cells forming late in the season (latewood) are smaller but have thicker walls—they provide strength instead of a means to transport sap.

The overall variation in wood fibers between earlywood and latewood determines whether the grain is even or uneven. In even-grain woods, there is little difference in density between the two; uneven-grain woods display a three- or fourfold difference in density. If you've ever stained a softwood, you've likely had problems getting a uniform color because of uneven grain—the lighter earlywood is more porous and will stain much darker than the latewood. Stain conditioners can help here.

3D softwood cell

If you could cut out a tiny cube of softwood and magnify it, it would look something like the bottom drawing. Note the radially aligned rays on both sides and the difference between earlywood and latewood. Here it's also easy to see how the wood fibers match up so the pits can align for fluid transfer.

HARDWOOD STRUCTURE

The cell structure of hardwoods is considerably more variable than that of softwoods—and that variability can produce spectacular effects. In particular, a combination of vessels that vary in diameter and configuration, along with wood fibers and rays, gives hardwoods the distinctive appearance that woodworkers covet.

Cells

Unlike softwoods, where two cells make up the majority of the wood's volume, most of a hardwood's volume is comprised of at least four types of cells. These are vessel elements, wood fibers, ray cells, and parenchyma or storage cells (*top drawing*). Some hardwoods have only 10% of the volume taken up by rays, with vessels and fibers taking up 60% and 30%, respectively. A different wood may be 35% rays, 50% fibers, and 15% vessels. Each species is different. Wood fibers in hardwoods tend to be shorter than in softwoods.

Vessels

Vessels occur in virtually all hardwoods, but *never* in softwoods. This is one of the easiest ways to identify a piece of wood as a hardwood. The configuration of the vessels has a huge impact on the wood's value to a woodworker—vessels affect its appearance, ability to dry, strength, finishing and machining, and more. Vessel elements are large in diameter and have relatively thin walls; they fit end to end, and fluid flows through perforations to form pathways along the length of a tree.

The vessels in some woods form tyloses as sapwood changes to heartwood. Tyloses are bubble-like membranes that develop inside the cavity of a vessel as it dies. The presence or absence of tyloses is what makes one wood impermeable to water and difficult to dry, but not another one. The vessels of white oak heartwood, for example, are clogged with numerous tyloses.

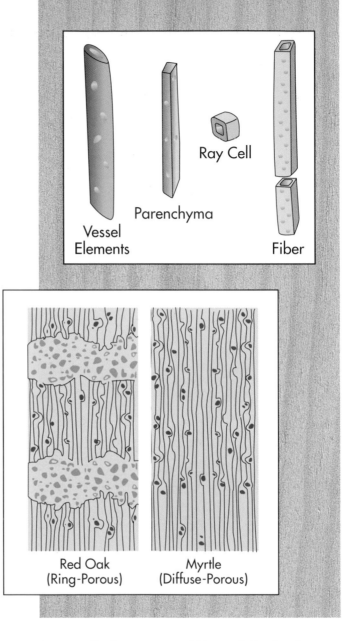

Vessel Elements   Parenchyma   Ray Cell   Fiber

Red Oak (Ring-Porous)   Myrtle (Diffuse-Porous)

Because fluid passage would be so difficult, it's used frequently to make "wet" casks for aging spirits such as wine and bourbon.

In hardwoods, pore size is used as a measure of texture. Woods with large pores, such as oak and ash, are termed coarse-textured. When pores are small, as in cherry or maple, the wood is said to be fine-textured. The porosity of a wood will greatly impact its ability to accept a stain or finish—much depends on whether it's ring- or diffuse-porous.

VESSEL VARIATIONS

White oak: Large vessels in white oak are distinct to the naked eye and are clustered in the earlywood, usually three pores in width.

Mahogany: Vessels are distinct to the naked eye and are numerous and evenly distributed. May have either white or gum deposits.

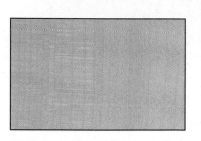

Sycamore: Vessels are uniformly distributed throughout and are small in earlywood and very small in latewood.

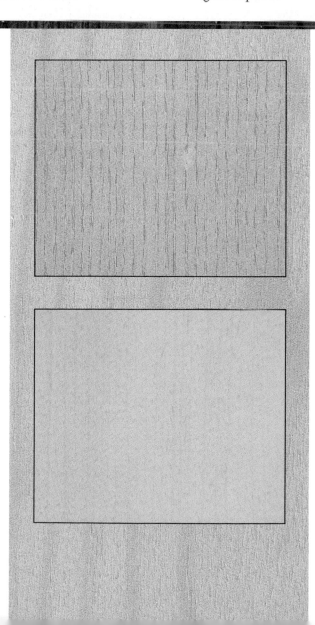

Ring-porous

Many hardwoods have high concentrations of vessels in their earlywood that are much larger in diameter than latewood vessels (*middle photo at left*). Species such as oak, elm, ash, and chestnut, which display distinct rings of very large vessels that can be seen with the naked eye, are called ring-porous hardwoods.

These woods usually have distinct figures and patterns. Although it's nice to look at, this variation in vessel size and concentration causes a pronounced uneven grain. Because of this, ring-porous hardwoods have a well-deserved reputation for uneven staining. The solution in most cases is to plug up the vessels by applying a paste wood "filler"; this results in a much more uniform color when stained.

Diffuse-porous

Diffuse-porous woods, where the vessels in the growth rings are all about the same diameter in both earlywood and latewood, are much easier to stain and finish (*bottom photo*). Most domestic diffuse-porous woods have relatively small pores, but tropical woods like mahogany can have larger pores.

Parenchyma

Hardwoods may or may not contain parenchyma, the specialized cells that are sort of a hybrid of a vessel and a wood fiber (*top drawing*). They serve primarily as storage cells and are very useful for wood identification.

Rays

Rays in hardwoods vary considerably in size and appearance. Rays provide a horizontal pathway for materials (particularly photosynthate) to flow from the inner bark toward the center of the tree. Rays are very important to woodworkers, not only for appearance but also because they create areas of structural weakness within the wood. As wood dries, internal stresses often create checks in these areas. Rays also allow firewood like oak to be split easily. On the downside, since rays are perpendicular to the fibers, they tend to chip out easily when machined.

But most of these traits can be overlooked because of the spectacular figure that rays produce in some woods. When white oak, sycamore, and lacewood are cut radially or quartersawn, they show large, shimmering ray fleck often referred to as silver grain (*see page 98 for more on this*).

3D hardwood cell

The middle drawing illustrates how all the cells in hardwood fit together. Note that the vessels run parallel and the rays run perpendicular to the wood fibers. Earlywood cells are large and thin-walled, and latewood cells are smaller, with thicker walls.

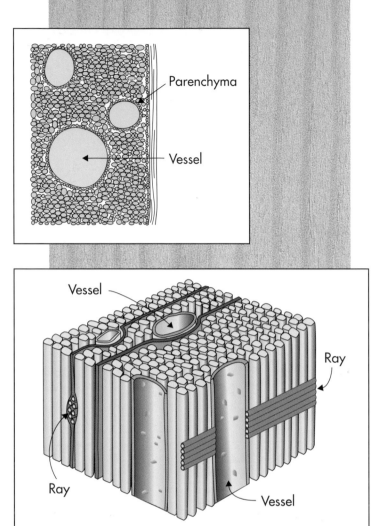

PROPERTIES OF HARDWOODS AND SOFTWOODS

	Hardwoods	Softwoods
Vessels	yes	no
Cell types	many	few
Cell composition	great variation	little variation
Rays	broad or narrow	narrow
Radial alignment of cells	no	yes
Growth ring visibility	can be difficult to distinguish	readily visible

KNOTS

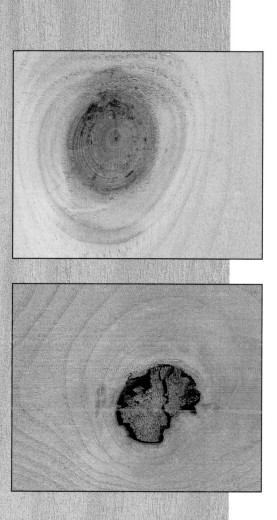

As a tree grows in height, branches develop from the pith out. Lateral branches are intergrown with the wood of the trunk as long as they are alive. That is, the cells of the tree and the branch intermingle. This type of knot is called a tight knot (*top photo*). When a branch dies, the trunk will continue to grow and surround it. This is known as a loose knot (*middle photo*). Eventually the branch will fall off and the stub will be overgrown until clear wood is formed once again.

Knots affect the strength of wood in two ways. First, the grain around the knot will be distorted. Its slope will cause a significant reduction in the strength of the wood around the knot. Second, if the knot is loose or encased, it contributes nothing to strength. The location of a knot on a board will determine how it affects strength. If it's near the edge, the strength of the board will be reduced according to the knot's size. When a knot is located near the center of the board, it has less impact on overall strength.

KNOTS: DEFECT OR ACCENT?

Woodworkers usually regard knots as defects—they cut them out and toss them onto the scrap pile. But in some woods, knots aren't considered a defect, but a desirable accent. As a matter of fact, the more there are, the better. One such wood is knotty pine (*shown here*). The knots in this pine are usually tight or sound, small and clustered. When used as paneling or in country pieces, they can add a rustic charm.

Another wood where knots are sought out is Eastern red cedar. Aromatic oils in the wood concentrate in the knots; more knots, better smell. Well-placed knots can also serve as a decorative accent on a piece; partial knots can even be used as a rustic pull for a door or drawer.

WOOD CHARACTERISTICS

Color. Texture. Grain. Figure. That's what comes to mind when most woodworkers are considering a wood for a project. But there are many more hidden characteristics to wood that can have as much impact, if not more, on the success or failure of a project. Every species of wood has a unique set of strength characteristics. For example, some woods bend easily, while others snap with the slightest deflection. Woods like ash and hickory stand up well to impact, but others like balsa and basswood crush easily.

Although there are many characteristics, I've described the more common ones here: specific gravity, modulus of rupture, modulus of elasticity, impact bending, compression parallel to grain, compression perpendicular to grain, and shear strength parallel to grain (*see the drawing on page* 21). For an exhaustive study of the many characteristics of wood, consult the *Wood Handbook: Wood as an Engineering Material,* published by the Forest Products Society (www.forestprod.org).

CHARACTERISTICS OF COMMON WOODS

	Specific gravity	Modulus of rupture (psi)	Modulus of elasticity (Mpsi)	Impact bending (in.)	Compression parallel to grain (psi)	Compression perpendicular to grain (psi)	Shear parallel to grain (psi)
Cherry, black	0.50	12,300	1.49	29	7,110	690	1,700
Douglas fir, coast	0.48	12,400	1.95	31	7,230	800	1,130
Hickory, shagbark	0.72	20,200	2.16	67	9,210	1,760	2,430
Maple, sugar	0.63	15,800	1.83	39	7,830	1,470	2,330
Oak, red northern	0.63	14,300	1.82	43	6,760	1,010	1,780
Oak, white	0.68	15,200	1.78	37	7,440	1,070	2,000
Pine, eastern white	0.35	8,600	1.24	18	4,800	440	900
Walnut, black	0.55	14,600	1.68	34	7,580	1,010	1,370

(all characteristics are for wood dried to 12% moisture content)

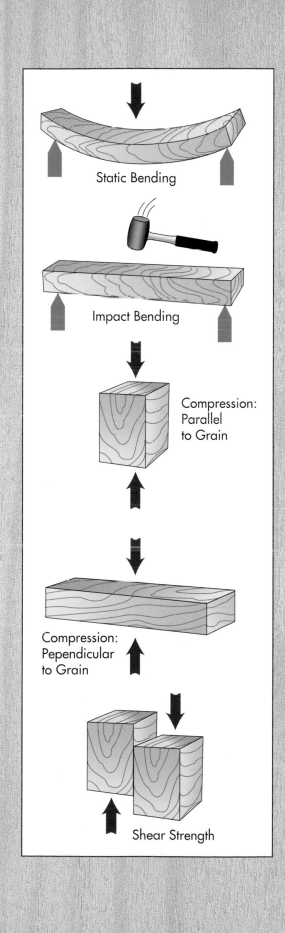

Static Bending

Impact Bending

Compression:
Parallel
to Grain

Compression:
Pependicular
to Grain

Shear Strength

Specific gravity

To compare the weights of different woods, specific gravity is used instead of actual weight because it's more accurate. Specific gravity is defined as "the ratio of the density of the wood to the density of the water." Specific gravity of wood is most often based on the oven-dry weight at a specified moisture content.

Modulus of rupture

You can compare the load-carrying ability of one wood to another by examining the modulus of rupture—the maximum load the wood can support without breaking. Say you want to make some bookshelves. What's better, eastern white pine or Douglas fir? The chart on *page 20* clearly shows that Douglas fir will support a lot more weight than pine.

Modulus of elasticity

The modulus of elasticity is a measure of a wood's ability to "spring back" after a load is removed. If you've ever missed hitting a nail and left a dent, you've encountered a nonelastic wood. The higher the modulus of elasticity, the better the wood will be able to recover. Wood for an archery bow needs high elasticity to return to its original shape after being drawn.

Impact bending

Impact bending describes how well a wood handles impact. In this test, a hammer is dropped from progressively greater heights until a rupture occurs or the wood deflects more than 6". Red alder is rated at 22", true hickory at 88"—it's no wonder that sports equipment is often made of hickory.

Compression and shear strength

Compression parallel to grain is how well a wood holds together when sustained stress is applied to the end grain. Compression perpendicular to grain identifies a wood's ability to hold up under stress applied to its surfaces. And finally, shear strength measures a wood's ability to resist internal slipping of one part along another, along the grain.

JUVENILE WOOD

The wood in every tree that forms within its first 10 years or so has different characteristics than the mature wood produced later in its life—most of it is undesirable. This wood is known as juvenile wood and can also be found at the tips of the main stem and branches. As long as a tree continues to grow in height, juvenile wood will be produced. But for the most part, juvenile wood occupies the center portion of every tree (*top drawing*).

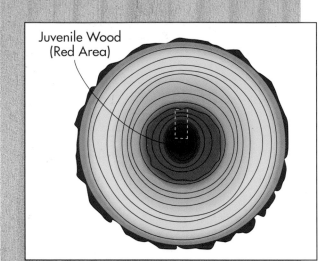

Juvenile Wood
(Red Area)

Properties of juvenile wood

Juvenile wood has two main characteristics that make it very undesirable to woodworkers. First, juvenile wood shrinks and swells both with the grain and along the grain. An example of this would be the 2×2 shown in the middle photo: Here, one-half of the piece is juvenile wood and its edge has shrunk in length much more than the other part, resulting in severe crook. Second, the strength of juvenile wood is much lower than the mature wood of the same tree—as much as 50% lower. This is because juvenile wood has more thin-walled earlywood and less thicker-walled latewood.

More common today

Juvenile wood is becoming more of a problem for woodworkers today because the average diameter of trees harvested has declined (*bottom photo*). In the past, when large logs were milled, the center section containing juvenile wood was set aside for pallet stock. Today, many mills will dip into the juvenile wood to eke out another board or two. Unfortunately, there is no way to clearly identify juvenile wood; your best defense is to steer clear of wood that contains the pith.

REACTION WOOD

When the trunk of a tree tips from vertical, it tries to bring itself back into the upright position by developing reaction wood (*top photo*). In softwoods, reaction wood forms on the underside of the leaning truck and is called compression wood; in hardwoods it develops on the upper side of the leaning truck and is termed tension wood. In either case, reaction wood has many of the undesirable qualities present in juvenile wood.

Identifying reaction wood

One of the biggest challenges of either type of reaction wood is identifying it. Unless you see the tree before it's harvested, it's difficult to tell whether reaction wood is present. Even the mill operator that receives the log will have a hard time knowing. There are some clues, however. In general, the growth rings in reaction wood are wider than in other parts. The area opposite the reaction wood generally has narrow growth rings. In softwoods, compression wood often appears reddish in color (*middle photo*). As a woodworker, the only way you can detect reaction wood is to study the end grain of the board. If you find unevenly spaced growth rings, chances are it contains reaction wood.

Fuzzy grain

In addition to shrinkage and strength problems, tension wood in hardwoods also creates another annoying problem: fuzzy grain (*bottom photo*). Loosely attached tension-wood fibers have a tendency not to cut cleanly, and portions of the wood fibers often tear out during machining. Sharp blades can help but will not make the problem go away. Your best bet is to write the board off and get a new one.

"We work with boards from these trees, to fulfill their yearning for a second life, to release their richness and beauty. From these planks we fashion objects useful to man, and if nature wills, things of beauty."

GEORGE NAKASHIMA (1981)

DIRECTORY of WOOD

We all share something with master furniture maker Nakashima: an appreciation of wood's function, beauty, and rich variety. From the inky gloss of ebony to the pale hues of aspen, wood pleases the eye, the hand, the soul of all who are part of its "second life."

I once showed my turned work at a juried art show, and the potter in the next booth was amazed at how different all the woods in my pieces were. He went from walnut bowl to zebrawood vase, caressing the wood and commenting on the beauty of each. Like that potter, we admire and appreciate wood. We study the color and observe the grain, rub our hand over it, even smell it. Sure, we build things, but it's the wood itself that we most enjoy.

This chapter examines 64 common woods that woodworkers use. Each piece is $1/2$" × 3" × 6"—the standard size used by the International Wood Collectors Society (IWCS). Founded in 1947, the IWCS is a nonprofit society devoted to collecting, crafting, and advancing information on wood. You can learn more about the IWCS at www.woodcollectors.org. To show both natural and finished states, each sample has been sanded with 150-grit sandpaper, and the lower half is finsihed with three coats of satin polyurethane.

Alder, Red

Alnus rubra

You may have walked on red alder if you've ever worn clogs, since it's the traditional wood used in this footwear. Found on the Pacific Coast of North America, this is a soft, weak wood. While it has low shock resistance and very low stiffness, it machines very well. Thanks to its uniform small pore structure and the absence of any visible boundary between heartwood and softwood, it accepts stains readily and takes a good finish. Red alder is one of the easiest commercial timbers to peel into veneer. Its natural defects—knots, burr or burl clusters, minor stains and streaks—are showcased in its decorative veneer form.

The sapwood is white to pinkish brown, and the heartwood is found only in red alders of advanced age. The wood has a subdued pattern and fine texture. It's used for furniture, veneer and plywood, sash and door panel stock, and other millwork.

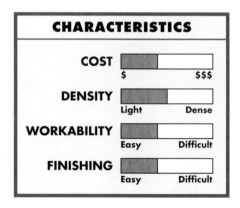

CHARACTERISTICS

- COST — $ — $$$
- DENSITY — Light — Dense
- WORKABILITY — Easy — Difficult
- FINISHING — Easy — Difficult

Ash, American

Fraxinus americana

At baseball and hockey games, ash is part of the action: Its high shock resistance makes it the preferred wood for baseball bats and hockey sticks, as well as for tool handles. This exceptionally flexible wood is found in eastern North America from Nova Scotia south to Georgia and west to the Mississippi River. Ash is often underappreciated by many woodworkers; it offers a sweet fragrance when cut, it's inexpensive, it machines well, and it can be stained to look very similar to red oak (*see page 179 for more on this*).

The sapwood of ash is nearly white; the heartwood is light brown to pale yellow. The grain is straight, coarse, and even-textured. Even though it's heavy, hard and strong, ash boasts remarkable bending properties. Coupled with its shock resistance, this flexibility makes it perfect for an array of uses: cabinetwork, veneers, bent handles for umbrellas, bent frames for canoes, boat oars, canoe paddles, and snowshoes.

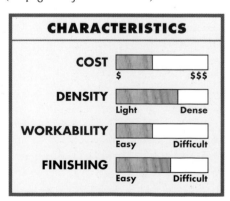

CHARACTERISTICS

- COST — $ — $$$
- DENSITY — Light — Dense
- WORKABILITY — Easy — Difficult
- FINISHING — Easy — Difficult

Aspen, American

Populus tremuloides

Aspen gets around: It covers a wide zone across the northern United States, and up into Canada. Famous for its flat, paper-thin leaves that flutter with even the slightest breeze, it's often called quaking aspen. Once aspen is dry, it has no odor or flavor, so it's well suited for use in the food industry as lightweight containers and cooking utensils. Wide, glued-up panels of aspen are becoming popular in home centers in the United States, but be aware that its woolly surface can cause problems in staining.

Aspen is whitish, creamy gray to gray-brown and has a straight to woolly pattern. The grain is a fine and even texture. Aspen is light and soft and often used for pulpwood—it is the preferred wood for making OSB (oriented-strand board). Aspen is also used to make plywood, furniture, and construction lumber. Due to its nondescript grain, it's often used as a paint-grade hardwood.

CHARACTERISTICS

COST	$ $$$
DENSITY	Light Dense
WORKABILITY	Easy Difficult
FINISHING	Easy Difficult

Balsa

Ochroma pyramidale

Balsa is a wood of extremes: It's the lightest-weight, commercially used wood in the world. It's also the softest and most porous wood (it's difficult to finish, as it absorbs whatever finish is applied)—yet it's a hardwood. What's more, this product of Central and South America and the West Indies grows extremely fast: It can be ready to harvest just six years from planting. It has excellent strength and stability despite its very light weight. Buoyant balsa (the word means "raft" in Spanish) is a good insulator against heat and cold; it's valued as a heat insulator in refrigerated ships.

Balsa is pale to pinkish white and has an indistinct pattern with some pores visible. Though associated with model planes, less than 10 percent of balsa is used for models and novelties. It's favored for floatation devices, life preservers, rafts, and speed boats, plus flooring in aircraft, recreational vehicles, and subway cars, where weight is a factor.

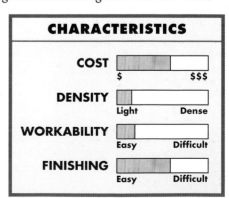

CHARACTERISTICS

COST	$ $$$
DENSITY	Light Dense
WORKABILITY	Easy Difficult
FINISHING	Easy Difficult

Basswood

Tilia americana

Carvers around the world are big users of basswood, also known as linden. The reason: Its grain offers both softness and regularity—lack of contrast between earlywood and latewood. Woodturners also account for basswood's popularity, especially when a project calls for lightness in weight: Turned Christmas-tree ornaments are often of basswood. While its fans are global, this wood's native ground is limited to the eastern half of North America and the Canadian provinces. Because it has no taste or odor at all, the business world uses, too: It's been used for years in the food industry for tools, utensils, and containers.

Basswood is creamy white in color with a fine, indistinct grain. A very light wood, it's fairly soft and weak. Those traits explain why it's used primarily for carving, hobbies and crafts, commercial veneer, food-handling utensils, and food containers. It's also used to manufacture beehives and Venetian blinds.

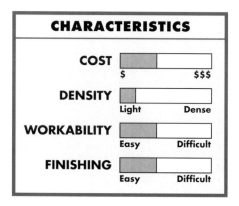

CHARACTERISTICS

COST	$ $$$
DENSITY	Light Dense
WORKABILITY	Easy Difficult
FINISHING	Easy Difficult

Beech, American

Fagus grandifolia

When a job calls for wood that's sturdy and can take a pounding—but still look good—beech can be the answer. Hard, strong, and stiff, yet suitable for steam-bending, beech is often used for not-so-fine furniture like school desks and chairs. Found in Nova Scotia and Prince Edward Island in Canada and throughout the eastern one-third of the United States, beech offers an odd group of properties. It's great for steam-bending while retaining its strength; it becomes slick with wear and so is perfect for drawer sides and runners; and it imparts no odor or taste, so it's used for food utensils and containers.

The heartwood of American beech is white to pinkish to reddish brown. It has conspicuous rays and tiny pores, a straight to interlocked pattern, and a close grain with a fine texture. This all-around general-purpose wood is also used to make handles and flooring.

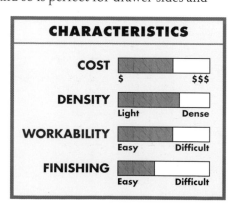

CHARACTERISTICS

COST	$ $$$
DENSITY	Light Dense
WORKABILITY	Easy Difficult
FINISHING	Easy Difficult

Birch, Yellow

Betula alleghaniensis

If you've ever bought high-quality plywood, it was probably yellow birch, since that's its biggest use. Abundant in the forests of the U.S. northeast and the Great Lakes states, this wood has a smooth, dense surface that's free from pores. It paints, stains, and polishes beautifully. Unequalled as a base for enameling, it virtually guarantees a permanent, smooth finish. Yellow birch steam-bends well and is a common choice in making upholstery frames.

The sapwood of yellow birch is white, while the heartwood is cream or light brown tinged with red. When stained, the difference between heartwood and sapwood is barely noticeable. The grain is straight, close, and even, and it has a fine texture. Yellow birch is heavy, hard, and strong. In addition to its primary use in plywood, it's also popular for furniture, interiors, and cabinetwork, and for door skins on flush doors and interior and exterior panel doors.

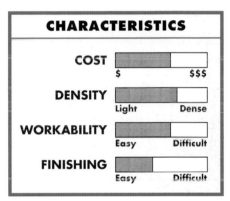

CHARACTERISTICS

COST	$	$$$
DENSITY	Light	Dense
WORKABILITY	Easy	Difficult
FINISHING	Easy	Difficult

Blackwood, Australian

Acacia melanoxylon

Australian blackwood, often referred to as black wattle, grows in New South Wales, Queensland, and southeastern Australia. There are hundreds of species of wattle belonging to the Acacia species, but Australian blackwood is the most attractive and largest growing. It is similar to ash in impact strength and has good steam-bending properties. Lustrous Australian blackwood is a highly decorative timber and is in great demand for furniture.

The sapwood of Australian blackwood is straw-colored, and the heartwood is not the color the name suggests; instead it is golden to dark reddish brown with occasional chocolate-colored splotches. The grain is medium-textured and straight but often interlocked and wavy, creating a handsome fiddleback figure. Heavy and strong, Australian blackwood is most used for furniture, cabinetwork, and gunstocks, and it can be sliced into beautiful, decorative veneer.

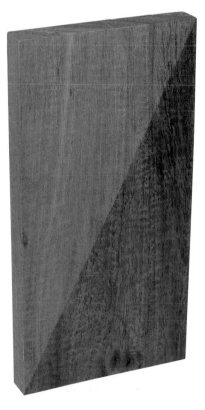

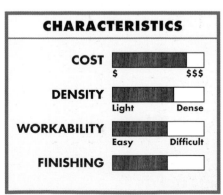

CHARACTERISTICS

COST	$	$$$
DENSITY	Light	Dense
WORKABILITY	Easy	Difficult
FINISHING		

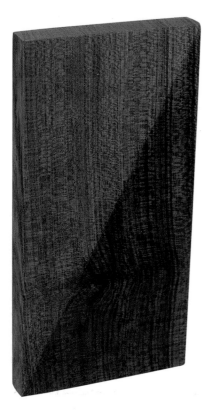

Bocote

Cordia spp.

If you like the look of teak—but don't like what it does to the cutting edges of tools—try bocote. Often used as a substitute for rosewood or teak, this wood is similar in texture and color to teak. Even though it's harder, it isn't as abrasive and will only slightly blunt cutting edges. Bocote is one of the many types of cordia found throughout the West Indies, tropical America, Africa, Asia, Mexico, Belize, and Honduras. It's typically available only in small sizes and is often sliced into decorative veneers.

The heartwood of bocote is rich, golden brown with a pinkish tint, often showing variegated irregular markings and an attractive ray fleck figure when quartersawn. It is straight-grained with a medium-coarse texture. Bocote is heavy and of medium strength. It is most often used in the manufacture of furniture, cabinetwork, boat decking, veneer, and tool handles.

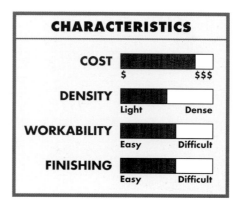

CHARACTERISTICS

COST		
	$	$$$
DENSITY		
	Light	Dense
WORKABILITY		
	Easy	Difficult
FINISHING		
	Easy	Difficult

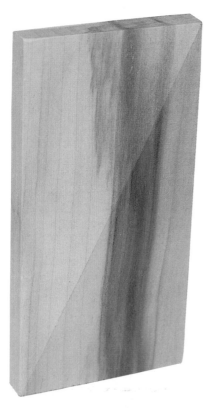

Box Elder

Acer negundo

Credit an unglamorous fungus for the beauty of box elder. That's what causes the red streaks that form stunning patterns in this wood. These impressive colorations make this member of the maple family sought-after by woodturners. Box elder is found in lower elevations in North America, extending through Mexico into Guatemala, excluding Pacific Coast states and south central Canada. While fast growing, it's also short-lived: This delicate tree is highly susceptible to damage from wind, heart rot, insects, and fungus.

The heartwood of box elder is yellowish brown, while the sapwood is greenish yellow to creamy white. Red streaks in the wood are composed of a pigment from a fungus. Because box elder is light, soft, porous, close-grained, and weak, it has limited cosmetic use. For the most part, it's used for inexpensive furniture, railroad cross-ties, woodenware, boxes, crates, pallets, and wood pulp.

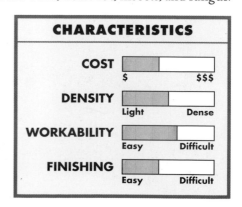

CHARACTERISTICS

COST		
	$	$$$
DENSITY		
	Light	Dense
WORKABILITY		
	Easy	Difficult
FINISHING		
	Easy	Difficult

Bubinga

Guibourtia demeusei

It looks like rosewood. It's often called "African rosewood." It's sometimes sold as a substitute for rosewood. But bubinga is a distinct wood with special properties. Found in Africa, primarily around Cameroon, Gabon, and Zaire, bubinga trees are massive, producing logs weighing up to 10 tons that produce extremely wide planks. When bubinga is rotary-cut into veneer to display its exotic coloring, it's sold under the trade name "Kevazingo." Bubinga causes moderate to severe tool blunting, and interlocked and irregular-grained areas tend to tear or pick up when machined. It needs a finish with UV protection to prevent fading to brown with exposure.

Bubinga is red-brown with light red to purple stripes or veining. It is straight-grained, often with irregular and interlocked areas. It is heavy, and gum pockets can cause problems machining and gluing. Bubinga is used for turnings, handles, furniture, and decorative veneer.

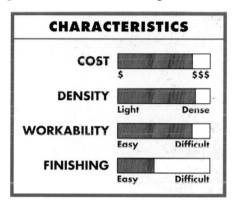

CHARACTERISTICS		
COST	$	$$$
DENSITY	Light	Dense
WORKABILITY	Easy	Difficult
FINISHING	Easy	Difficult

Butternut

Juglans cinerea

There's something about this wood that appeals to both carvers and clerics. Its natural luster comes up quickly when polished, and it's the wood most often used for church altars. Found in the United States from Maine south to Virginia and west to Iowa and Missouri, it's also known as white walnut because it resembles black walnut but is lighter in color. Although it can be stained to look like black walnut, it doesn't offer the same strength or hardness. Easy to work with, butternut is an excellent carving wood. Virtually every part of a butternut tree is sticky—the leaves, the stems, the flowers, the seeds, the nuts, and the sap are all oily and buttery to the touch.

The heartwood of butternut is warm, medium brown. It is straight-grained and coarse with a soft texture. Butternut is light to medium weight and is used for furniture, carving, interiors, and cabinetwork and is sliced into decorative veneer.

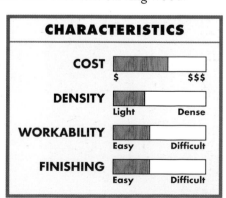

CHARACTERISTICS		
COST	$	$$$
DENSITY	Light	Dense
WORKABILITY	Easy	Difficult
FINISHING	Easy	Difficult

Cedar, Eastern Red

Juniperus virginiana

While its reputation for moth-repelling is unproven, Eastern red cedar's pleasant scent is undisputed: It was one of the first woods to be exported from America to Europe, where perfume was extracted from its oils. If it's the scent you're after, look for knots: More knots indicate higher concentration of scent-producing oils. Eastern red cedar grows in North America from Nova Scotia south to Georgia along the Atlantic coast and west to the Mississippi River. Expect a lot of waste working with this wood because of loose knots and problems matching boards due to the highly varied colors.

The sapwood of eastern red cedar is nearly white; the heartwood is purplish to rose red, which matures to dull red or reddish brown. The grain is fairly straight with a fine texture. Eastern red cedar is moderately heavy and hard. It is used for chests, wardrobes, closet interiors, paneling, drawer lining, and pencils.

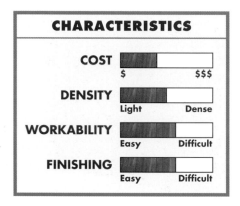

CHARACTERISTICS		
COST	$	$$$
DENSITY	Light	Dense
WORKABILITY	Easy	Difficult
FINISHING	Easy	Difficult

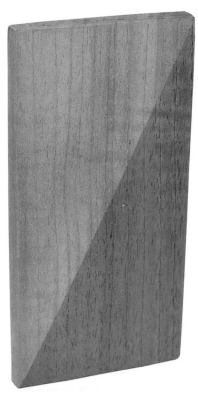

Cedar, Spanish

Cedrela odorata

A true cedar (*Cedrus* spp.) is a softwood, but Spanish cedar is a hardwood. It's a common misnomer: Many hardwoods that produce a fragrance similar to softwood cedar are called cedar, and this is one. It grows in Mexico and South America. Spanish cedar contains gum that gives the wood its characteristic pleasant odor. This gum may exude, and appears on the surface as sticky resin, which can lead to difficulties in finishing. Spanish cedar is often referred to as cigar box cedar. It's known as the wood of choice for making cigar humidors.

Spanish cedar is pale pinkish brown to dark reddish brown. The grain is straight but may be wavy, curly, or mottled in areas—it can exhibit a moderate to high golden luster. It is moderately light and soft. Spanish cedar is used for cigar boxes, pencils, and doors and is sliced into decorative veneer for paneling and plywood.

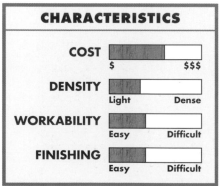

CHARACTERISTICS		
COST	$	$$$
DENSITY	Light	Dense
WORKABILITY	Easy	Difficult
FINISHING	Easy	Difficult

Cedar, Western Red

Thuja plicata

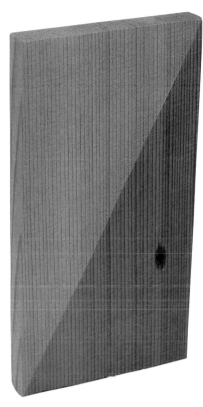

Water attacks both houses and boats, and the best defense is often western red cedar. Completely nonresinous, this is one of the most decay-resistant species in America, hence its popularity in home and marine construction. It grows along the coastal ranges of western Canada and the United States from Alaska through British Columbia, Washington, and Oregon and east to Montana. Because of its fragrance, it's often confused with "aromatic" cedar. Its tendency to split makes it perfect for shingles; when exposed, it weathers to an attractive silver gray.

The sapwood is white; the heartwood is a rich red color. It is straight and even-grained with a fine pattern and texture. Western red cedar is moderately soft and light in weight, is extremely decay-resistant, and exhibits little shrinkage. Low in strength and brittle, it's used primarily for shingles and shakes, siding, caskets, boatbuilding, porch columns, and sheds.

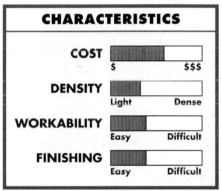

CHARACTERISTICS

COST	$ — $$$
DENSITY	Light — Dense
WORKABILITY	Easy — Difficult
FINISHING	Easy — Difficult

Chakte Kok

Sickingia salvadorensis

You could say that chakte kok (pronounced CHOCK-tay COKE) is chock-full of color. In fact, the wood and bark of this tree are used to create a commercial red dye. Chakte kok grows in the Yucatan and Chiapais regions of southeastern Mexico and Belize and is often available from sustainably managed sources. The waste factor can be high when working this wood because it's difficult to match boards—they can vary widely in color. Chakte kok requires a finish with UV protection; otherwise it will quickly fade when exposed to sunlight.

The heartwood of chakte kok is a vibrant pinkish red with occasional streaks of brown. It often possesses a gorgeous flame figure. The grain is fine and even, with occasional check and wormholes. Chakte kok is hard and dense. It is used for furniture and cabinetwork and is sliced into decorative veneers for paneling and plywood.

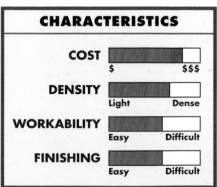

CHARACTERISTICS

COST	$ — $$$
DENSITY	Light — Dense
WORKABILITY	Easy — Difficult
FINISHING	Easy — Difficult

Chechem

Metopium brownei

The chechem (pronounced chay-CHEM) tree is one to look at—not to touch. The bark contains a caustic sap that can raise a poison ivy–like rash on skin. The wood, though, is eminently touchable, with a hardness and density like Brazilian rosewood. In color, chechem resembles black walnut with a golden luster, making it look like teak (in fact, it's a good substitute for teak). Chechem grows in the Dominican Republic, Cuba, Jamaica, Guatemala, Belize, and Mexico. It is often available from sustainably managed sources.

When finishing, know that chechem takes lacquer finishes well, but not polyurethanes.

The heartwood of chechem is shaded amber to dark brown, often with a range of colors and contrasting streaks. It is tight-grained, hard, and dense. Chechen is slightly oily and very rot-resistant. It is used for furniture making, cabinetwork, and boat decking and is sliced into veneer for paneling or plywood.

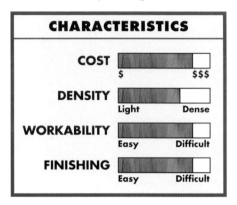

CHARACTERISTICS		
COST	$	$$$
DENSITY	Light	Dense
WORKABILITY	Easy	Difficult
FINISHING	Easy	Difficult

Cherry, Black

Prunus serotina

One of the most highly prized American woods for cabinetmaking, cherry gleams with advantages: It cuts cleanly, polishes well, and turns a gorgeous, rich red as it ages. Once dried, it's so stable that it's used for end-grain engravers' blocks and to back the metal engravers' blocks used to print U.S. currency. Cherry grows in North America from the Canadian border south to the Carolinas and west to the Dakotas. Although it machines well, it has a well-deserved reputation for burning if you hesitate during a cut.

The sapwood is nearly white; the heartwood, light pinkish brown. The variance in color between heartwood and sapwood can be problematic—this discrepancy will become more obvious as the heartwood darkens with age. The grain is straight with a fine, close pattern. It is light, strong, and hard. Cherry is used for turning, carving, furniture, interiors, cabinetwork, musical instruments, and decorative veneer.

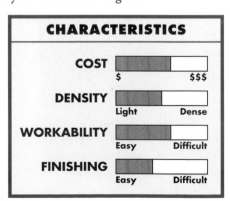

CHARACTERISTICS		
COST	$	$$$
DENSITY	Light	Dense
WORKABILITY	Easy	Difficult
FINISHING	Easy	Difficult

Chestnut, American

Castanea dentata

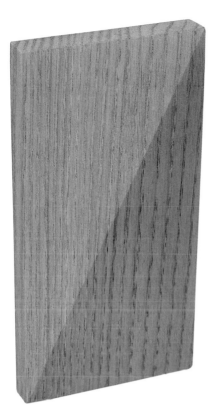

American chestnut once made up one-third of the U. S. hardwood growth (about 9 million acres of forest). In the 20th century, a fungus bark disease nearly made this species extinct. All current supplies of chestnut lumber come from standing dead timber located in "ghost forests." Although you can still can find logs in the Appalachian Mountains (since the fungus attacks the bark and not the wood), this source is almost depleted. New chestnut trees continue to sprout, but they are quickly killed by the bark disease. American chestnut is often referred to as wormy chestnut because it is liberally marked with wormholes.

The heartwood of chestnut is straw to light brown in color. The grain is fairly straight and coarse textured, sprinkled with wormholes. Chestnut is medium weight and strong. It is used locally for furniture and is sliced into veneers for paneling and plywood.

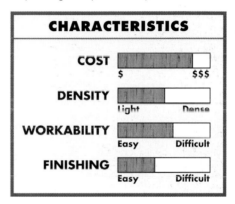

CHARACTERISTICS

COST		$ ——— $$$
DENSITY		Light ——— Dense
WORKABILITY		Easy ——— Difficult
FINISHING		Easy ——— Difficult

Cypress, Bald

Taxodium distichum

Where moisture is a challenge, bald cypress can be the solution. Found in North America from New Jersey south throughout the southeastern United States, this is a strange tree in many ways. It's very long-lived and thrives in swamp water, preferring to be submerged at least part of the year. Although it's a conifer, it loses its needles every winter. In the spring it produces new, tender green needles; in the fall, they turn autumn colors and fall off again. Bald cypress boasts extremely high resistance to insects, dampness, and fungal decay. So it's excellent in any locale where insects and humidity are a concern.

This wood is yellowish red to salmon-colored. It has a distinct leafy grain and can exhibit a nice crotch pattern. It is moderately strong, reasonably light, and durable. Bald cypress is used for furniture, interior and exterior trim, tanks, vats, greenhouses, stadiums, and barrel making.

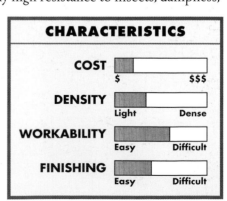

CHARACTERISTICS

COST		$ ——— $$$
DENSITY		Light ——— Dense
WORKABILITY		Easy ——— Difficult
FINISHING		Easy ——— Difficult

Douglas Fir

Pseudotsuga menziesii

Named for David Douglas, the Scottish botanist who discovered it, Douglas fir is the forester's idea of perfection—very tall with a thick trunk tapering up gradually with no branches for the first 100 feet. Douglas fir grows along the Pacific Coast of North America from British Columbia through Washington and Oregon into California, and eastward into the Rocky Mountains. It is one of the most widely used woods in North America and the continent's most plentiful species: The vast stands in Washington and Oregon carry more usable timber per acre than any other forest except redwood. Although not a true fir, it's a fast grower and is very long-lived (500 years is common).

Douglas fir is pale cream to light orange in color, straight-grained,. and hard and strong—exceptionally strong for a softwood. It is used primarily for construction lumber; it is the world's most important source for plywood.

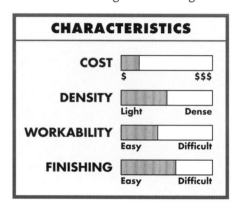

CHARACTERISTICS		
COST	$	$$$
DENSITY	Light	Dense
WORKABILITY	Easy	Difficult
FINISHING	Easy	Difficult

Ebony

Diospyros spp.

For exotic pasts, it's hard to beat this very hard wood. Ebony was the ultimate wood for the ancient Greeks, the maharajahs of India, and the pharaohs of Egypt. Its black heartwood has been treasured for centuries. Ebony grows in the Philippines, East Indies, India, Sri Lanka, and Madagascar as well as Africa, Central America, and South America. Extremely expensive, today ebony is most often used as an accent such as fingerboards on violins and keys on pianos. It tends to be very brittle, and the wood is so dense it will dull virtually any tool. The sawdust from ebony has been known to cause respiratory problems.

The heartwood of ebony is an uneven gray to black with black stripes to jet black. The grain is slightly interlocked with a fine texture. Ebony is very heavy. It is used for musical instrument parts and turnings, is sliced into decorative veneer, and is often made into inlay.

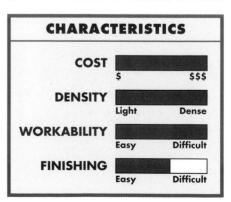

CHARACTERISTICS		
COST	$	$$$
DENSITY	Light	Dense
WORKABILITY	Easy	Difficult
FINISHING	Easy	Difficult

Elm, American White

Ulmus americana

Though relatively hardy, the American white elm couldn't stand up to Dutch elm disease. That's why no native American elm grows west of the Rocky Mountains, a region that used to boast hundreds of thousands of these trees. They still grow in North America, from Newfoundland to Nova Scotia in Canada south to Florida along the Eastern seaboard. The American white elm has excellent steam-bending and water-resistant properties. Even perpetual wetness didn't faze the elms of old: Romans hollowed out elm trees and used them for water pipes.

The sapwood of American white elm is grayish white; the heartwood is light-brown with a reddish tinge. The grain is straight, sometimes interlocked with a coarse texture. It is moderately heavy, and weak. Elm is a standby in dam and lock construction; due to its muddy color, it's most often used as a paint-grade hardwood for furniture and cabinetwork.

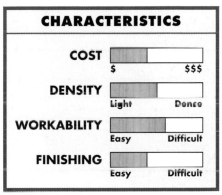

CHARACTERISTICS

COST		
	$	$$$
DENSITY		
	Light	Dense
WORKABILITY		
	Easy	Difficult
FINISHING		
	Easy	Difficult

Gonçalo Alves

Astronium fraxinifolium

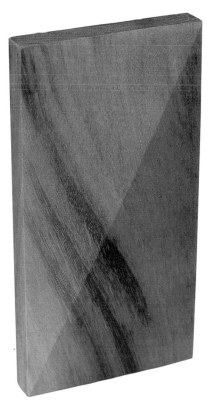

Brazil yields this handsome, rewarding—but challenging—wood. Gonçalo alves (pronounced gon-SALL-o ALL-vez) turns well and finishes smoothly, with a vibrant russet gleam. But since the heartwood is highly resistant to moisture absorption, it can be difficult to glue. And gonçalo alves takes extra effort to work because it blunts cutting edges. It's called zebrawood in the United Kingdom and tigerwood in the United States.

The heartwood of gonçalo alves is russet to orange brown, with narrow to wide stripes of medium to very dark brown, contrasting sharply with the brownish white sapwood. It has an irregular grain, often interlocked with alternating layers of hard and soft material with an overall medium texture. Gonçalo alves is heavy, hard, and dense. Although it is used as an accent and for high-quality furniture and cabinetwork and for turnings, it is available primarily in decorative veneers.

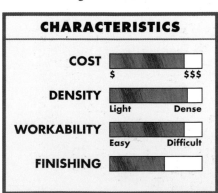

CHARACTERISTICS

COST		
	$	$$$
DENSITY		
	Light	Dense
WORKABILITY		
	Easy	Difficult
FINISHING		

Gum, Sweet

Liquidambar styraciflua

Nice smell...nasty movement. That's the good news/bad news about sweet gum, found in North America from New England to Mexico and into South America. It produces vanilla-scented resin formed in the bark by wound stimulation. The resin is a source of storax or styrax, used in medicine and perfumery. Nice. But it tends to warp and twist when dried, requiring great care in drying to avoid degradation. Plus, it exhibits a lot of wood movement after it's been dried. Nasty. Sweet gum's heartwood is sold separately as red gum, and the sapwood is sold separately as sap gum.

The heartwood of sweet gum is a dull pinkish brown; the sapwood is creamy white. The grain is irregular but with a fine texture. Sweet gum is moderately heavy and hard but is not exceptionally strong. It is used for turnings, cabinetwork, decorative and commercial veneer, and sometimes for "dry" or slack barrel making.

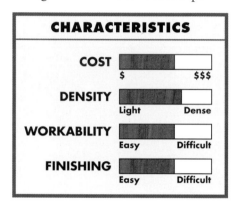

CHARACTERISTICS

COST	$	$$$
DENSITY	Light	Dense
WORKABILITY	Easy	Difficult
FINISHING	Easy	Difficult

Hickory

Carya spp.

When you need a wood with flexibility, strength, and resilience, consider hickory. With high bending strength and crushing strength, high stiffness, and very high shock resistance, its outstanding combination of properties makes hickory great for striking tools like hammers and axes, and also for sports equipment like bats and golf clubs. Hickory grows in North America, from southeastern Canada to the eastern half of the United States. The heartwood is sold as red hickory; the sapwood is sold as white hickory. This wood shrinks considerably when drying but then is stable. It causes moderate blunting of cutting edges.

The sapwood of hickory is nearly white; the heartwood is creamy to pinkish brown. The grain is straight, close, and finely textured. It is moderately hard and heavy, extremely tough, and resilient. Hickory is used for tool handles, ladder rungs, sports equipment, furniture, turnings, and cabinetwork.

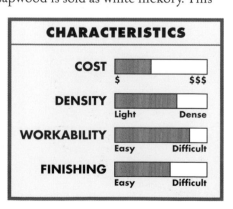

CHARACTERISTICS

COST	$	$$$
DENSITY	Light	Dense
WORKABILITY	Easy	Difficult
FINISHING	Easy	Difficult

Holly

Ilex spp.

This is a wood that can change identities. Since holly can be virtually white with no visible figure, it's sometimes used as a substitute for boxwood. At the other extreme, it's often dyed black to mimic ebony. There are over 100 species, found in Europe, the United States, and western Asia. It is excellent for turning. Because very little holly is cut each year and the trees are small, it is used mostly for inlay. Holly is difficult to dry and tends to split. For best results, it should be converted to small pieces and then dried slowly with the top of the pile weighted down.

The heartwood of holly is white to ivory white with bluish streaks. The grain is even with no visible figure and has a fine texture. Holly is hard and moderately strong. It is used for turnings, for musical instrument parts, for accents in furniture and cabinetwork, and often as inlay.

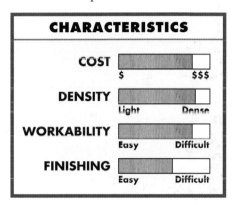

CHARACTERISTICS		
COST	$	$$$
DENSITY	Light	Dense
WORKABILITY	Easy	Difficult
FINISHING	Easy	Difficult

Imbuya

Octea porosa

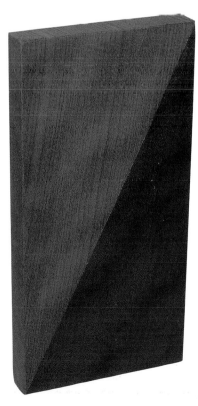

If you admire a piece of furniture in Brazil and are told it's Brazilian walnut, you're probably looking at this wood. Imbuya grows in Brazil and can look so much like black walnut that it is often referred to as Brazilian walnut. Although imbuya is not as strong as black walnut, it is highly valued in its native country as a cabinet and furniture wood. Imbuya has a peculiar, spicy odor, most of which is lost in drying. The sawdust often causes irritation.

The heartwood of imbuya is rich brown with some streaks and stripes. The grain is fairly straight, often wavy or curly in areas, with a fine texture. It is fairly hard, heavy, and durable, making it a good choice for furniture and cabinetry. It is used primarily for decorative purposes and is commonly seen as veneer or in paneling, or as rifle butts and gunstocks.

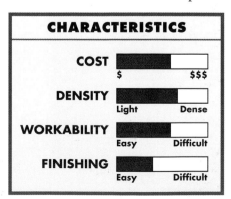

CHARACTERISTICS		
COST	$	$$$
DENSITY	Light	Dense
WORKABILITY	Easy	Difficult
FINISHING	Easy	Difficult

Iroko

Milicia excelsa

Carvers, turners, and boatbuilders share a liking for this wood from equatorial Africa. Iroko is sometimes referred to as African teak, but it isn't as good-looking as teak and doesn't have teak's greasy feel. It's a favorite wood for carving and turning, and like teak, it's widely used in boatbuilding. Iroko causes moderate to severe blunting effect on cutting tools. If you value your cutting tools and your sinuses, approach this wood with caution. It is not popular in North America, because the logs often contain hard deposits of calcium carbonate, commonly called "stone," which make them difficult to work; these can severely damage tools. The sawdust of iroko can cause respiratory problems.

The heartwood of iroko is light brown to rich golden-orange brown. The grain is interlocked, with a coarse texture. Iroko is moderately heavy, hard, and dense. It is used for boatbuilding, furniture, and cabinetwork and is sliced into veneer.

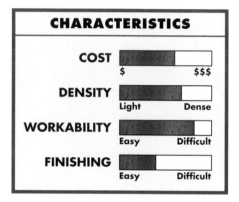

Jarrah

Eucalyptus marginata

This type of eucalyptus has a special status in its native land: It's harvested more than any other Australian timber. Jarrah is used locally in Australia for projects with a maritime theme: shipbuilding and marine structures like dock pilings and harbor work, and bridge building. Jarrah requires careful drying, due to its tendency to twist and warp. Gum pockets or veins are a common defect that can make it difficult to work.

The heartwood of jarrah is dark brownish red and is often marked with short, brown radial flecks on the end grain and boat-shaped flecks on the flat-sawn surfaces. Fungus is the cause of these flecks, which enhance the wood's decorative value. The grain is straight to irregular, often wavy with a coarse texture. Jarrah is heavy and of medium strength. It is used for furniture, boatbuilding, marine construction, tool handles, and cabinetwork and is sliced into decorative veneer.

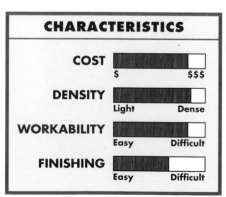

Jatoba

Hymenaea courbaril

Shock-resistance similar to that of ash is the prime trait of this wood, which grows in Central and South America and the West Indies. It is often available from sustainably managed sources. While it's a good choice for tool handles and sporting goods, it can be difficult to machine because of its high density; it causes moderate to severe blunting of tool edges. Jatoba tree bark contains an orange or yellowish resin called South American copal, used in the manufacture of specialty varnishes and cements. Jatoba is difficult to dry and often exhibits surface checking.

The heartwood of Jatoba is salmon-red to orange-brown, marked with dark brown to russet brown streaks. It often exhibits a golden luster. The grain is inter-locked and of a coarse texture. Jatoba is heavy, hard, and tough and has good bending characteristics. It is used for furniture, cabinetwork, sporting goods, and tool handles and is sliced into decorative veneer.

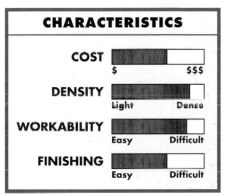

CHARACTERISTICS		
COST	$	$$$
DENSITY	Light	Dense
WORKABILITY	Easy	Difficult
FINISHING	Easy	Difficult

Kingwood

Dalbergia cearensis

It's easy to see why this wood is so named. A member of the rosewood family, kingwood grows in Brazil, and its unmistakable appearance makes it a king among woods. It was heavily used in the finest furniture for Louis XIV and Louis XV of France and in the Georgian period of English furniture. That's why it's extremely popular now with antiques restorers. The great de-mand for antiques restoration work has made this an endangered species, and it's available only in small pieces. It causes moderate dulling of cutting edges; but if tools are kept sharp, a very smooth finish is obtainable—it is well suited to a wax finish.

The heartwood of kingwood is violet-brown to black with dark streaks of violet-brown, black, and sometimes golden yellow, often lus-trous. The grain is straight, with a fine texture. Kingwood is heavy, strong, and tough. It is used mostly for accent work, marquetry, inlays, and turnings.

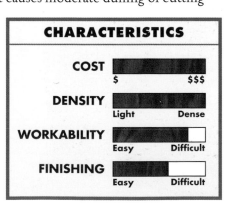

CHARACTERISTICS		
COST	$	$$$
DENSITY	Light	Dense
WORKABILITY	Easy	Difficult
FINISHING	Easy	Difficult

Koa

Acacia koa

There's a musical connection with this wood, which grows only in the Hawaiian Islands and is the principal timber of the 50th state. Koa is best known for helping make music in ukuleles and guitars, but it is also used for gunstocks. As supplies dwindle, this ever-scarcer wood is now used chiefly to produce highly decorative veneer. Its fiddleback figure lends it well to architectural paneling and decorative face veneering. Cutting edges must be kept sharp, especially when curly grain is encountered.

The heartwood of koa is golden reddish brown, with dark brown streaks marking the growth rings, showing as black lines on longitudinal surfaces. The grain is interlocked, often with a wavy or fiddleback pattern. It is medium-textured and is sometimes lustrous. Koa is moderately heavy and hard. It is used for furniture, musical instruments, and gunstocks and is sliced into veneer for paneling and plywood.

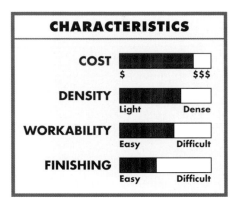

CHARACTERISTICS

COST	$	$$$
DENSITY	Light	Dense
WORKABILITY	Easy	Difficult
FINISHING	Easy	Difficult

Lacewood

Cardwellia sublimis

There's no mistaking the look of lacewood, which grows in Australia and is often referred to as silky oak. Lacewood and silky oak are two related but separate species. Lacewood has a much more obvious lustrous quartersawn figure—large rays produce a silver grain figure much more pronounced than in silky oak (*see page* 47). Neither species, though, is a true oak (*Quercus* spp.). Quartersawn lacewood has a high tendency to chip out when planed, and the sawdust is an irritant and can cause respiratory problems.

The heartwood of lacewood is light pink with a silvery pink sheen. The grain is straight with a coarse texture; ray fleck is prominent when quarter-sliced. Lacewood is moderately heavy and hard. In Australia, it's often a substitute for softwood in building and shuttering, interiors, and so forth. Elsewhere it is used for furniture, inlays, accents, and turnings and is often sliced into decorative veneer.

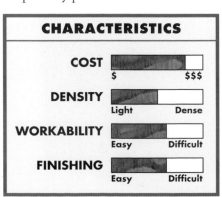

CHARACTERISTICS

COST	$	$$$
DENSITY	Light	Dense
WORKABILITY	Easy	Difficult
FINISHING	Easy	Difficult

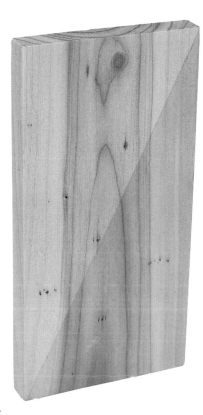

Larch, European

Larix decidua

For a softwood, European larch is very tough. Used where durability and strength are prime requirements, this tree grows in Europe, particularly the mountain areas of the Alps, in the United Kingdom, and in western Russia. Although it's a conifer, it sheds its needles in the winter. It is often treated chemically and used outdoors for stakes, transmission poles, boat planking, bridge construction, railway sleepers, and exterior joinery that contacts the ground. European larch is resinous and must be seasoned properly to avoid problems. The knots can cause moderate to severe blunting of cutting edges.

The sapwood of European larch is white; the heartwood is light orange-red. The grain is straight and of uniform texture. It's durable and tough and boasts abrasion properties superior to many softwoods. European larch is commonly used for staircases, flooring, window frames, posts, and fencing

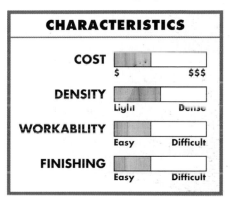

CHARACTERISTICS

COST		
	$	$$$
DENSITY		
	Light	Dense
WORKABILITY		
	Easy	Difficult
FINISHING		
	Easy	Difficult

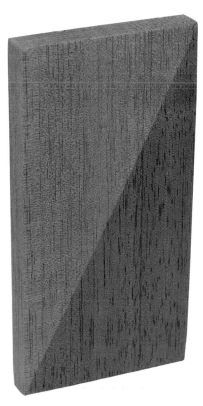

Lauan

Shorea spp.

A common nickname for this wood is "Philippine mahogany," but luckily for woodworkers' wallets, that isn't accurate. Lauan (pronounced lu-ON) is not actually a mahogany: It's much less rare, and therefore much less expensive. It's also known as the working man's plywood because it is used for subflooring as well as backs, bottoms, and drawers in furniture making, and as a substrate in paneling and plywood. It can be found in southeast Asia, from the Philippine Islands south and west throughout Indonesia. Lauan has excellent waterproof properties and is often used in boatbuilding.

The heartwood of lauan is medium to dark reddish brown, with whitish resin streaks. The grain is interlocked, with a coarse texture. It is moderately heavy and hard. Lauan is used for cabinetwork, for interiors, for doors, for commercial and decorative veneer, and in the manufacture of plywood.

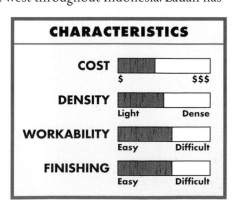

CHARACTERISTICS

COST		
	$	$$$
DENSITY		
	Light	Dense
WORKABILITY		
	Easy	Difficult
FINISHING		
	Easy	Difficult

Locust, Honey

Gleditsia triacanthos

Although it grows in several regions of the United States, this sturdy wood is fairly scarce. Honey locust grows in North America from Pennsylvania west to South Dakota, Nebraska, south to Texas, east to Alabama and Georgia and northeast along the Appalachians. The "Gleditsia" in its botanical name is a Latinized word, honoring Johann Gottleib Gleditsh (1717–1786), director of the Berlin Botanic Garden. Honey locust has many desirable qualities: attractive figure and color, hardness and strength, and no odor or taste; but is not commonly used, because it's in short supply.

The sapwood of honey locust is creamy white; the heartwood is pinkish to reddish brown. The grain is straight and the wood is heavy and hard. Honey locust is used locally for fence posts and rails and general construction. Occasionally it is used for crafts, for furniture, and sliced into veneer.

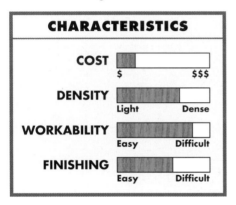

CHARACTERISTICS

COST	$ — $$$
DENSITY	Light — Dense
WORKABILITY	Easy — Difficult
FINISHING	Easy — Difficult

Mahogany, African

Khaya spp.

African mahogany grows in West, Central, and East Africa. It is closest to American mahogany or "true" mahogany (*Swietenia mahogoni*)—the premier wood for fine furniture and cabinetwork in Europe as early as the 1600s. Today, African mahogany has a solid reputation as a quality wood. It's an important timber for furniture, office desks, cabinetwork, staircases, banisters, handrails, and boatbuilding. In maritime construction, it's extensively used in laminations, especially in cold-molded process.

The heartwood of African mahogany varies from light to deep reddish brown. The grain is straight to interlocked, with a medium to moderately coarse texture. It produces a nice roe figure when quartersawn. African mahogany is moderately heavy and hard. It's used for furniture and cabinetwork and is sliced into decorative veneer for paneling and plywood.

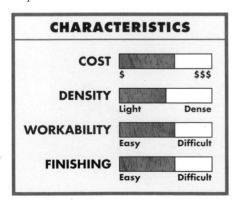

CHARACTERISTICS

COST	$ — $$$
DENSITY	Light — Dense
WORKABILITY	Easy — Difficult
FINISHING	Easy — Difficult

Mahogany, Big Leaf

Swietenia macrophylla

Often referred to as "New World" mahogany, this variety can be found in Mexico south through Central America to Brazil and Peru. It rates among the best furniture woods in the world. Big leaf mahogany stains and polishes to a beautiful luster. The trees are huge, and the lumber is available in many widths, lengths, and thicknesses. It is prized for its figures: stripe, roe, curly, blister, fiddleback, and mottle. Big leaf mahogany works extremely well and is the wood of choice for pattern makers: New cars are often made entirely out of this mahogany—every part—then the parts are used as patterns.

The heartwood of big leaf mahogany varies from light to dark reddish brown to deep red. The grain is straight to interlocked, and medium to moderately coarse in texture. It can be highly figured when sliced into veneer. Moderately hard and heavy, this wood is used for furniture, cabinetwork, decorative veneer, and musical instruments.

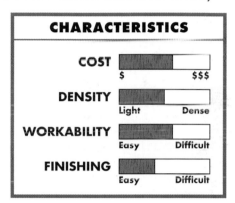

CHARACTERISTICS

COST	$ — $$$
DENSITY	Light — Dense
WORKABILITY	Easy — Difficult
FINISHING	Easy — Difficult

Maple, Hard

Acer saccharum

Hard maple (also known as rock maple or sugar maple) grows in North America in the Rocky Mountains of Canada and the eastern states of America. It is famous for exhibiting rare figure in some trees: wavy, curly, quilted, blistered, fiddleback, burl, and the much sought-after bird's-eye pattern. Its unusual resistance to abrasion and indentation makes it the number one choice for furniture and flooring—especially dance floors and bowling alleys. This tough wood has a tendency to burn during end-grain cuts.

The sapwood of hard maple is white to creamy white; the heartwood is creamy white to pinkish tinge to light reddish brown. The grain is straight, sometimes wavy or curly, with a fine texture. It can be highly figured. Hard maple is heavy, hard, and tough. It's used for furniture, flooring, interiors, cabinetwork, decorative veneer, woodenware, bowling pins, spools, and handles.

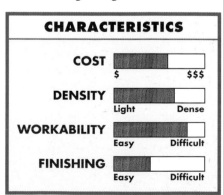

CHARACTERISTICS

COST	$ — $$$
DENSITY	Light — Dense
WORKABILITY	Easy — Difficult
FINISHING	Easy — Difficult

Maple, Soft

Acer rubrum

Soft maple can be found in North America in north temperate regions of Canada and the eastern United States and along the Pacific Coast. It resembles hard maple but is not as heavy and is much easier to work. The wood of silver maple, red maple, and box elder is soft maple; that of sugar maple and black maple is known as hard maple.

The sapwood in soft maple is white; the heartwood is gray-white to pinkish tinge to light reddish brown, often with olive or greenish black areas that are known as mineral streak, which is likely due to an injury. The sapwood in soft maples is considerably wider than that in hard maples. The grain is straight, with a fine texture. Soft maple is heavy and fairly tough. It is used for furniture, furniture parts, and cabinetwork. Specialized uses include shoe lasts, dairy equipment, sporting goods, and musical instruments.

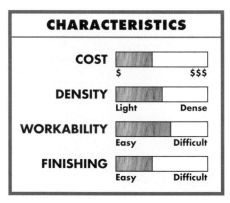

CHARACTERISTICS

- COST — $ $$$
- DENSITY — Light | Dense
- WORKABILITY — Easy | Difficult
- FINISHING — Easy | Difficult

Oak, American Red

Quercus rubra

American red oak is hugely popular in the United States for interior trim and furniture making. No wonder: It works exceptionally well, but because of its open pores, it must be filled properly before finishing. It can be found in North America in eastern Canada and the eastern half of the United States. Its high shock-resistance and crushing strength make it ideal for flooring. When quartersawn, American red oak produces a nice ray fleck. It also exhibits excellent steam-bending qualities.

The sapwood of American red oak is grayish white to pale reddish brown; the heartwood is pinkish to light reddish brown. The grain is straight with a coarse texture. It is heavy, hard, and strong. American red oak is used for furniture, cabinetwork, interior trim, and slack or "dry" barrel making and is often sliced into veneer for paneling and plywood.

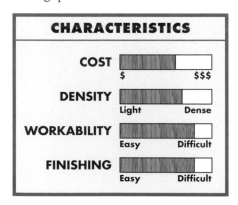

CHARACTERISTICS

- COST — $ $$$
- DENSITY — Light | Dense
- WORKABILITY — Easy | Difficult
- FINISHING — Easy | Difficult

Oak, American White

Quercus alba

What makes this wood significantly more valuable than red oak? Not only is it very resistant to wear, insects, and fungi, but it's also practically waterproof. American white oak is found along the eastern half of the United States. Its medium bending and crushing strengths combined with its low stiffness make it an excellent choice for steam-bending. The sap-conducting pores of American white oak are naturally plugged, yielding its water repellency. It's used for "wet" or tight casks and is especially prized for aging and storing wine, bourbon, and whiskey.

The sapwood is whitish to light brown; the heartwood is rich light brown to dark brown. The grain is straight, with a coarse texture, and shows a prominent ray fleck when quartersawn. American white oak is heavy, hard, and strong. It's used for furniture, cabinetwork, flooring, tight or "wet" barrel-making, keels, planking, and bent parts in ships and boats and is sliced into veneer.

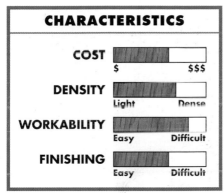

CHARACTERISTICS		
COST	$	$$$
DENSITY	Light	Dense
WORKABILITY	Easy	Difficult
FINISHING	Easy	Difficult

Oak, Silky

Grevillea robusta

Silky oak is a name given to a number of different genera and species in Australia and New Zealand. Although lacewood (*see page* 42) is often referred to as Australian silky oak, the name originally referred to the oak shown here, native to southern Queensland, which also grows in New South Wales. Silky oak is similar to lacewood but doesn't have as pronounced a figure when quartered. It is drought-resistant and is reported to have been successfully grown in warm, dry regions throughout the world, including the southeastern United States. Contrary to what its name suggests, the sawdust of this wood can cause skin irritation.

The sapwood of silky oak is cream or straw-colored; the heartwood is pale or red-brown upon drying. The grain is typically straight, sometimes wavy, and often lustrous. It is moderately heavy and strong and is used for furniture and inlays and can be sliced into decorative veneer.

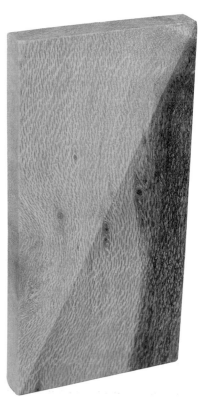

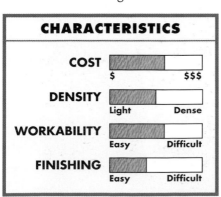

CHARACTERISTICS		
COST	$	$$$
DENSITY	Light	Dense
WORKABILITY	Easy	Difficult
FINISHING	Easy	Difficult

Osage Orange

Maclura pomifera

This tree went to war in the early 20th century: Its bark was used in the United States to make khaki dye for uniforms worn during World War I. Osage orange can be found in North America, tropical America, and Africa. The genus Maclura is dedicated to William Maclure (1763–1840), an American geologist. It is exceptionally resistant to decay and is one of the most durable woods in North America. It is commonly confused with black locust.

The sapwood of osage orange is light yellow; the heartwood is golden yellow to bright orange, darkening upon exposure to sunlight. The grain is straight, with a medium to fine texture. Osage orange is heavy, hard, strong, and durable. It was used by Native Americans to make archery bows and is used now for turnings, fence posts, smoking pipes, crutches, insulator pins, wheel rims and hubs of farm wagons, and machinery parts.

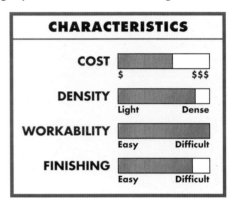

CHARACTERISTICS		
COST	$	$$$
DENSITY	Light	Dense
WORKABILITY	Easy	Difficult
FINISHING	Easy	Difficult

Padauk, African

Pterocarpus soyauxii

Vivid in color, high in strength, durable, and decay-resistant, African paduak is a workhorse. Known as camwood or barwood in the United Kingdom, it grows in central and tropical West Africa. This dramatic-looking wood is often used in boatbuilding, and its abrasion-resistance makes it ideal for flooring. African padauk is also a world-famous dye-wood. The timber works well with only moderate blunting (the sawdust will stain your hands), but needs UV protection or it will darken quickly.

The heartwood of African padauk is a vivid blood red, changing to dark purple-brown with red streaks upon exposure; the sapwood is straw-colored. Its grain is straight to interlocked, with a coarse texture. African padauk is heavy, hard, and strong. In some parts of the world, it is commonly used as flooring. Elsewhere it's used for furniture, cabinetwork, turning, accents, and inlays and is sliced into decorative veneer.

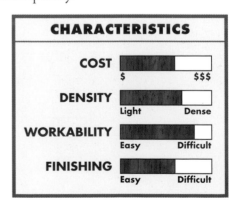

CHARACTERISTICS		
COST	$	$$$
DENSITY	Light	Dense
WORKABILITY	Easy	Difficult
FINISHING	Easy	Difficult

Pear

Pyrus communis

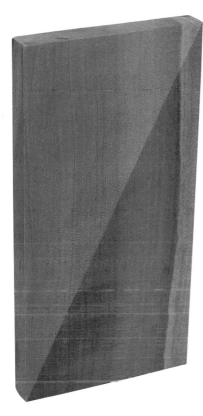

Although the fruit of the tree is valued more than the wood, pear has a very fine, even texture that makes it great for carving. It grows in Europe, the United States, and western Asia. Since the small tree limits the size of lumber, it makes excellent turning stock, and in Europe it is often used to make recorders. Pear can also be dyed black to resemble ebony. It has a tendency to warp when drying and causes a blunting effect on the cutting edges of tools.

The sapwood of pear is pale yellow apricot; the heartwood is a pale pinkish brown. The grain is fairly straight, with a smooth, even texture. Rays are visible on quartersawn surfaces as deeper flecks of a darker tone. Pear is heavy, hard, and moderately strong. It is used for carving, turning, musical instruments, cabinetwork, and decorative veneer for paneling, marquetry, and inlay.

CHARACTERISTICS		
COST		
	$	$$$
DENSITY		
	Light	Dense
WORKABILITY		
	Easy	Difficult
FINISHING		
	Easy	Difficult

Pecan

Carya illinoinensis

Undervalued by many woodworkers, this is a fine, attractive wood. Pecan is prized more for its nut crop than its timber and is often confused with hickory (a cousin). It's found in central and southeastern United States, most often in the flood plains of the Mississippi River and other low-lying terrain. The high bending strength and crushing strength, high stiffness, and very high shock-resistance of pecan make it an excellent choice for steam-bending. The tree has a short trunk and many forked branches, which makes it rather difficult to obtain long lengths of lumber.

The heartwood of pecan is reddish brown; the sapwood is white. The grain is straight, although often mottled or wavy with a coarse texture. It is used for turnings, furniture, tool handles, sports equipment, and drumsticks, and it can be sliced into veneer.

CHARACTERISTICS		
COST		
	$	$$$
DENSITY		
	Light	Dense
WORKABILITY		
	Easy	Difficult
FINISHING		
	Easy	Difficult

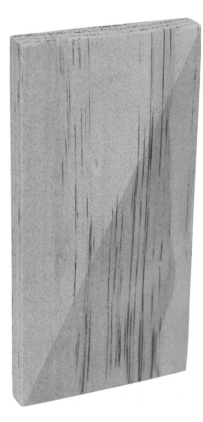

Pine, Eastern White

Pinus strobus

This wood helped start a war. The first sea skirmish of the American Revolution was over this lumber, which grows in Newfoundland, Ontario, and Manitoba in Canada, and throughout New England and the Great Lakes region to South Carolina in the United States. Eastern white pine was used so extensively for ship masts in the 18th and 19th centuries that the British government tried to impose laws reserving large trees, but the laws were openly flouted. The result was that historic sea clash. In the 18th and 19th centuries, virtually every colonial American building was constructed with eastern white pine—inside and out.

The sapwood of eastern white pine is nearly white to pale white; the heartwood is creamy to light to reddish brown. The grain is straight with a fine texture. Eastern white pine is light and moderately strong. It is used for construction lumber, patterns and casting, furniture, cabinetwork, and interiors.

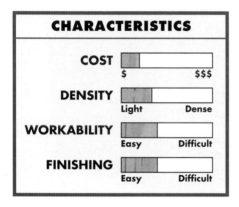

CHARACTERISTICS

COST	$ $$$
DENSITY	Light Dense
WORKABILITY	Easy Difficult
FINISHING	Easy Difficult

Pine, Ponderosa

Pinus ponderosa

Often marketed as "knotty pine" for interior decoration, this is one of the most attractive pines. Since the knots are usually sound, logs can be sliced into knotty pine veneer for paneling. Ponderosa pine can be found in western Canada and the western United States. With this most resinous of Canadian pines, resin exudation is the chief problem: The wood needs to be treated carefully before painting or varnishing. The durable sapwood of the ponderosa pine is also valuable and is used for pattern making.

The sapwood is light yellow; the heartwood is darker yellow to reddish brown. The heartwood is considerably heavier than the softwood, which is soft, uniform in texture, and nonresinous. The grain is straight, with an even texture. Ponderosa pine is used primarily in construction and for interior trim, turning, doors, window frames, boxes and packing, and furniture.

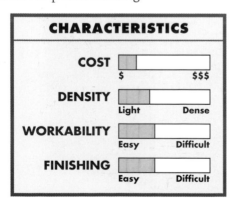

CHARACTERISTICS

COST	$ $$$
DENSITY	Light Dense
WORKABILITY	Easy Difficult
FINISHING	Easy Difficult

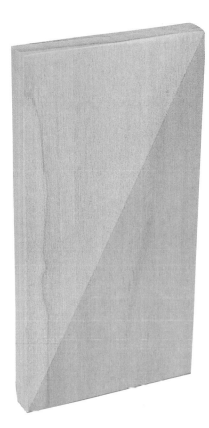

Poplar, Yellow

Liriodendron tulipifera

With no limbs or branches except at the very top, yellow poplar lumber has few knots. Often referred to as tulip tree, it grows in the eastern United States, from Connecticut south to Florida and west to Missouri. A fast grower, its lumber is lightweight and soft for a hardwood. Poplar paints well and can be stained to resemble walnut or cherry. It is gaining in popularity as a substitute for clear pine in interior trim.

The sapwood is nearly white; the heartwood is yellow to tan to greenish brown, frequently marked with streaks of purple, dark green, blue, or black. The grain is straight, with a fine to medium texture. Poplar is moderately light and soft. It is used for furniture, turnings, cabinet-work, interiors, and commercial veneer. Plentiful and affordable, poplar is also used for inexpensive products like toys, broom handles, baskets, food containers, Popsicle sticks, and tongue depressors.

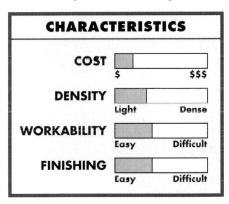

CHARACTERISTICS		
COST	$	$$$
DENSITY	Light	Dense
WORKABILITY	Easy	Difficult
FINISHING	Easy	Difficult

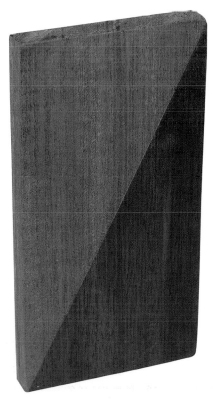

Purpleheart

Peltogyne paniculata

Its royal hue is temporary: Purpleheart is brown when cut and turns purple when exposed to air. But without UV protection, it will revert to brown when exposed to sunlight. Also called amaranth, it grows in Central and South America but mostly in the Brazilian Amazon region. There it's used in cabinetry and furniture, but also for flooring and heavy outdoor construction work like bridges, pilings, docks, and harbor work. Cutting can be hampered by gum deposits. If blades aren't kept extremely sharp, gummy resin often exudes when heated by a dull blade. Spirit finishes tend to remove the purple color; lacquer finishes tend to preserve the color

The heartwood of purpleheart is deep purple violet, maturing to dark brown. The grain is straight and often irregular and wavy, with a fine to moderate texture. Purple-heart is heavy and hard. It is used for furniture, cabinetwork, and inlays and is sliced into decorative veneer.

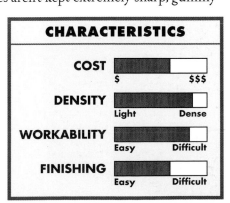

CHARACTERISTICS		
COST	$	$$$
DENSITY	Light	Dense
WORKABILITY	Easy	Difficult
FINISHING	Easy	Difficult

Redwood, California

Sequoia sempervirens

What makes this lumber a legend? The ability of California redwood to grow super fast, along with its decay- and insect-resistance and great longevity, makes it the most commercially valuable softwood. California redwood grows along the coastal northwest of the United States. Related to the Giant Sequoia (which are protected), the California redwood is often huge: 14 feet in diameter, 40 feet at the lower trunk, and 275 feet tall. Today most redwood for commercial growth comes from new-growth forest in privately held lands. The cinnamon-colored, very thick bark is used in the manufacture of fiberboard.

The sapwood of California redwood is nearly white; the heartwood is light red to deep reddish brown. The grain is straight, with a coarse texture. California redwood is light to moderately light and soft. It is used for outdoor furniture, decking, and siding and is sliced into decorative veneer.

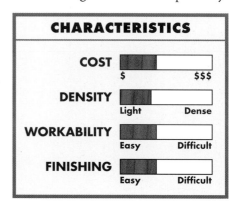

CHARACTERISTICS

COST	$ $$$
DENSITY	Light Dense
WORKABILITY	Easy Difficult
FINISHING	Easy Difficult

Rosewood, Brazilian

Dalbergia nigra

Exquisite color and exotic patterning…strikingly wild, wavy markings…all make this irregular pattern almost impossible to "match." Brazilian rosewood grows in the tropical zones of Mexico, Central America, and South America. It gets its name not from its color but from its odor—a fragrant oil in the heartwood has the scent of rose blossoms. Though one of the world's most treasured woods, it is increasingly rare due to a lack of a good forestry programs, past exploitation, and strict export embargoes. Brazilian rosewood has a severe blunting effect on tools.

The heartwood is chocolate-brown to violet-brown to violet, streaked with black or golden brown. The grain is straight to wavy, with a coarse texture. It is heavy, hard, and very dense. Brazilian rosewood is used in musical instruments for fingerboards and piano cases, for furniture, inlays, and decorative veneer, and often as a design element in knife and tool handles.

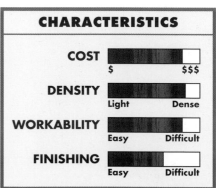

CHARACTERISTICS

COST	$ $$$
DENSITY	Light Dense
WORKABILITY	Easy Difficult
FINISHING	Easy Difficult

Sassafras

Sassafras albidum

Yes, it's more than tea, which is made from its root bark. Sassafras grows in North America from Maine through Ontario, Michigan, Iowa, and Kansas, to Florida and Texas. It is a member of the same family as cinnamon and is best known for its fragrant oil, used for flavoring and scenting, and of course, the tea. Sir Walter Raleigh brought sassafras back to England from the Virginia colony. It is seldom available in large sizes and is often mixed in and confused with black ash. Sassafras is decay-resistant and so is used for boatbuilding and making paddles.

The sapwood of sassafras is light yellow; the heartwood is dull grayish brown to darkish brown, sometimes with a slight greenish cast. The grain is straight, with a medium texture. Sassafras is moderately heavy and hard. It is used for furniture, for furniture parts, for cabinetwork, and in boatbuilding.

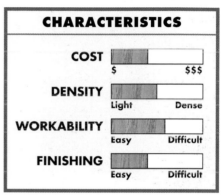

Satinwood, Ceylon

Chloroxylon swietenia

Ever since the "Golden Age of Satinwood" in 19th century England, this timber has been highly valued and in great demand. Although it has been used for centuries by cabinetmakers, it is scarce now and is found mainly in veneer form—especially the famous bee's-wing mottle pattern. Satinwood grows in central and southern India and Sri Lanka. Select logs are sliced to produce extremely attractive veneers in a wide variety of figures for paneling. Satinwood is lustrous and smells great, almost edible—like honey.

The heartwood of Ceylon satinwood is golden yellow, maturing to golden brown with darker streaks. There is little distinction between heartwood and sapwood. The grain is interlocked, with a fine texture. When quartersawn, it yields beautiful mottled and ribbon-stripped figure. Ceylon satinwood is heavy and hard. It is used for furniture, cabinetwork, accents, and inlays and is sliced into decorative veneer.

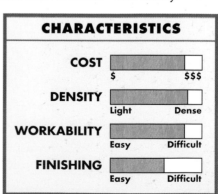

Snakewood

Brosimum guianense

You have to really love this wood to work it. One of the most expensive woods in the world, it's a small, rare timber with markings resembling snakeskin, sometimes spotty like leopard spots, or even like hieroglyphics. The last is why snakewood is called letterwood in the United Kingdom. It grows in Central and South America, from the Amazon region of Brazil through Colombia, southern Mexico, and the West Indies. The dark spots are the result of variations in the gummy deposits that fill the cell cavities. Snakewood is a challenge for more reasons: It needs care in gluing and finishing because of its resin, and it causes severe blunting of cutting tools.

The heartwood is reddish brown with snakeskin markings. The grain is interlocked, but moderately fine-textured. Snakewood is heavy, dense, and hard. It's used for turnings, carvings, musical instrument parts, and archery bows and is sliced into decorative veneer.

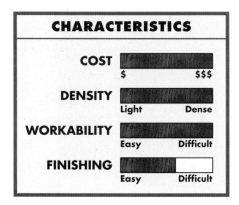

CHARACTERISTICS

COST	$	$$$
DENSITY	Light	Dense
WORKABILITY	Easy	Difficult
FINISHING	Easy	Difficult

Spruce, Sitka

Picea sitchensis

Read all about it! Spruce is the world's most important pulp for newsprint, because of its whiteness. And it's by far the most important wood for aircraft construction (remember Howard Hughes' *The Spruce Goose?*). Spruce grows in North America, from western Canada down to California. It is the major timber-producing tree of the Pacific Northwest, where its biggest use is in framing and general construction. Special grades of spruce are selected for their resonance to make soundboards for pianos, and guitar and violin fronts. This special spruce is typically quartersawn and often aged.

Spruce is white to yellowish brown with a slight pinkish tinge. The grain is straight, with a medium even texture. It has a very high strength-to-weight ratio. Spruce is used for general construction, millwork and interior trim, musical instruments, boatbuilding, oars, gliders, wood pulp, and plywood.

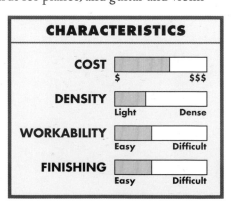

CHARACTERISTICS

COST	$	$$$
DENSITY	Light	Dense
WORKABILITY	Easy	Difficult
FINISHING	Easy	Difficult

Teak

Tectona grandis

Teak is so heavy that elephants are still used to transport big logs from inland jungles of southern India, Thailand, Myanmar (Burma), Java, and the East Indies to the waterways. It is distinctively oily to the touch; oil in the grain makes it very durable. Teak is resistant to all insects, fungus, and marine borers; termites won't touch it. And it resists rot and moisture. This vaunted wood is extremely abrasive and causes severe blunting to cutting tool edges. In fact, savvy teak turners keep an oilstone next to their lathes so they can sharpen their turning tools frequently. Even the leaves are abrasive: They're used locally as sandpaper. Teak sawdust can be an irritant.

Teak's heartwood is golden brown with dark chocolate streaks. The grain is straight, sometimes interlocked or wavy, and coarse-textured. Teak is moderately heavy and hard and oily. It's used for boat decks, flooring, "brightwork," and furniture and is sliced into veneer.

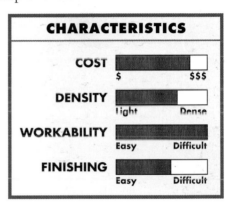

Walnut, Black

Juglans nigra

Beautiful black walnut has always been prized as a custom furniture and cabinetry wood. It can be highly figured: crotches, burls, fiddleback, butts, and stripes. Found in North America, black walnut is exceptionally stable once seasoned, and takes a high polish when finished. Walnut is famed as the "gunsmith's" wood: In addition to its beauty, its straight grain will not warp, guaranteeing accuracy. Black walnut is toxic to other plants—a mature tree produces a substance called juglone in its roots that kills neighboring trees. It is also toxic to animals: Its shavings should never be used for bedding.

The sapwood is whitish to yellowish brown; the heartwood is light gray-brown to rich chocolate-brown to deep purplish brown. The grain is straight to interlocked or curly or wavy, with a medium to coarse texture. Black walnut is moderately heavy, hard, and strong. It is used for furniture, cabinetwork, and carvings and is sliced into veneer.

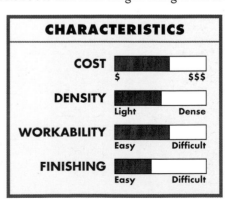

Wenge

Millettia laurentii

Like many other exotics, wenge is difficult to work but delightful to look at. This dark wood can be found in Equatorial Africa, particularly Zaire, Cameroon, and Gabon. Its high abrasion resistance makes it a choice wood for flooring in buildings or rooms (like conference rooms and hotel lobbies) where a high impact is expected. Wenge is excellent for turning, but its coarse grain needs to be filled for best appearance. It causes moderate blunting of cutting tools and can be difficult to glue because of its resins.

The sapwood of wenge is whitish; the heartwood is dark brown, with fine almost-black veins. The grain is usually straight, with a coarse texture. Wenge is heavy, hard, and dense. It is used for cabinetwork, furniture, accents, and inlays and is sliced into decorative veneer.

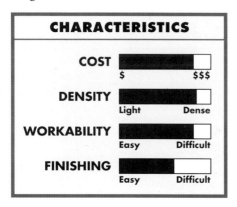

CHARACTERISTICS		
COST	$	$$$
DENSITY	Light	Dense
WORKABILITY	Easy	Difficult
FINISHING	Easy	Difficult

Willow, Black

Salix nigra

Because its European cousin is used to make cricket bats, the overseas relative is called "cricket willow." But this North American version grows in Canada, and in the eastern United States from the Atlantic Coast west to Minnesota, Iowa, Nebraska, Kansas, Oklahoma, and Texas, and south into Mexico. It is light, resilient, and flexible and has no characteristic odor or taste. Black willow's strength plus lightness makes it suited for use in artificial limbs. Black willow requires sharp cutting tools to prevent fuzzy or woolly grain.

The sapwood of black willow is light tan; the heartwood is light brown to pale reddish to grayish brown, frequently with darker streaks. The grain is straight, with a fine texture. Black willow is light and moderately soft. It is used most often for commercial veneer, wooden toys, and artificial limbs.

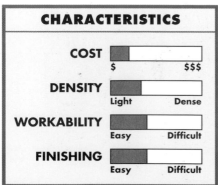

CHARACTERISTICS		
COST	$	$$$
DENSITY	Light	Dense
WORKABILITY	Easy	Difficult
FINISHING	Easy	Difficult

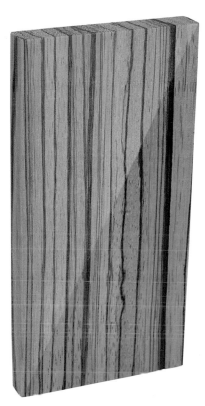

Zebrawood

Microberlinia brazzavillensis

Zebrawood looks incredible, but it's difficult to work. It tends to split, it's difficult to obtain a smooth finish because of the nature of the grain, and it smells (the first time you work it, you'll check your shoes to make sure you didn't step in something). Zebrawood's coarse texture requires a clear filler to get a smooth finish. Also called zebrano, it grows in West Africa, particularly in the Cameroons and Gabon. Its distinctive grain pattern comes from quartersawing. It is usually found as veneer or used as an accent. Note: The veneer tends to buckle over time and so should be weighted when it's being stored.

The heartwood is light golden yellow with dark brown to almost-black veins; the zebra stripe results from quartersawing. The grain is interlocked to wavy, with a coarse texture, but often lustrous. Zebrawood is heavy and hard. It is used for turning, accents, inlays, and marquetry and is sliced into decorative veneer.

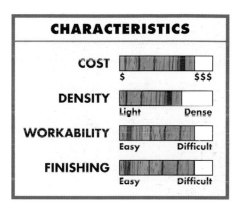

CHARACTERISTICS		
COST	$	$$$
DENSITY	Light	Dense
WORKABILITY	Easy	Difficult
FINISHING	Easy	Difficult

Zircote

Cordia dodecandra

If you can't find wenge or even ebony, try zircote. Found in Belize and in Mexico, this moderately hard wood is a member of the Cordia family—like bocote (*see page* 30). In fact, this wood has a lot in common with bocote, differing mainly in color. (Actually, you may often find the names interchanged at even the best lumberyards.) Zircote is a different species than bocote, and it's quite a bit darker in color. Most woodworkers find it relatively easy to work and turn for a variety of furniture projects.

The heartwood of zircote is black, gray, or dark brown with black streaks. The grain is straight, with medium to moderately fine texture, often lustrous. Zircote is moderately heavy and dense. It is used for furniture, cabinetwork, inlays, and accents and is sliced into decorative veneer.

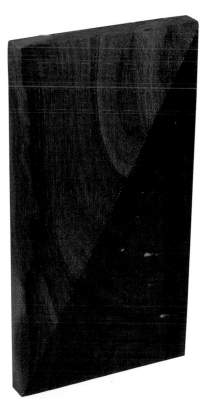

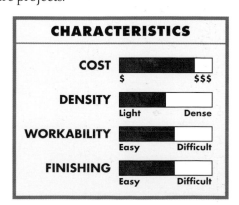

CHARACTERISTICS		
COST	$	$$$
DENSITY	Light	Dense
WORKABILITY	Easy	Difficult
FINISHING	Easy	Difficult

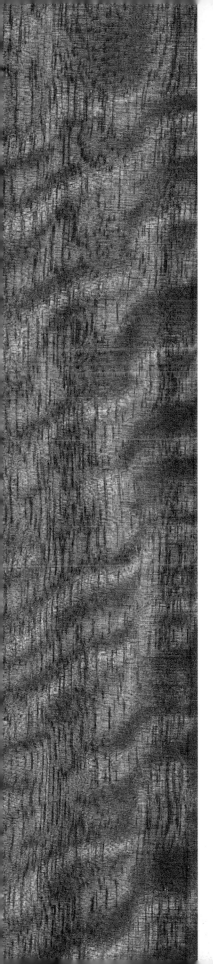

> "Before a log is opened, let the inquirer carefully examine it, and, noting all the peculiarities of its appearance, endeavor to form an opinion of its qualities; the event as seen in the cut wood will shew the value of the opinion he had formed."
>
> BLACKIE AND SON (1853)

MILLING LUMBER

Sawmills make me feel like a kid at Christmas. Every log is like a present, and you just can't tell from the wrapping what treasure might be inside. Sometimes, you find junk (Socks. Oh, thanks). But other times, as the smell of freshly cut wood spices the air, you lay open gorgeous, shimmering grain...beautiful ray fleck...wild spalting.

With wood, understanding how these potential treasures are revealed by the blade will help you to be a better woodworker. Once you know the different ways wood is cut, you can use this knowledge to create works that are finer, more lasting, and more beautiful.

Actually, there's no reason not to do your own cutting; it adds an extra dimension to your work. There's something especially rewarding about taking a project from the real start to finish—from tree to final result. This gives you a sense of completion that you don't get with store-bought lumber. Consider, too, what a family treasure you might create from a tree with special significance, like the maple Grandpa planted at his first house. Whether it's chainsaw milling, cutting logs on a bandsaw, or harvesting bowl blanks from a storm-damaged tree, many woodworkers find cutting just plain fun.

Typical Cuts

There are three main techniques that sawyers use to cut a log up into hardwood lumber: a through-and-through or "flitch" cut, quartersawing or rift-sawing, and plain-sawing or "sawing for grade" *(top to bottom in drawing at right)*. The difference among the three is how the growth rings of the log are oriented to the cut.

When a log is cut through and through, the lumber is cut tangentially to the growth rings. With quartersawing, the log is divided into quarters or thirds and then each section is cut radially to the growth rings. If a log is sawed for grade (the most common cut in hardwood mills), it's rotated as it's cut to yield the best lumber; this produces flat-sawn lumber along with rift- and quarter-sawn. The low-quality pith is usually set aside for pallet stock.

How can you tell how a board was cut? Just inspect the end grain. On a plain- or flat-sawn board, the annual growth rings will be 30° (or less) to the face of the board. A rift-sawn board will have growth rings that are more than 30° but less than 60°; with quartersawn boards, the growth rings will be 60° to 90° to the face of the board. The closer the rings are to 90°, the more stable the wood will be.

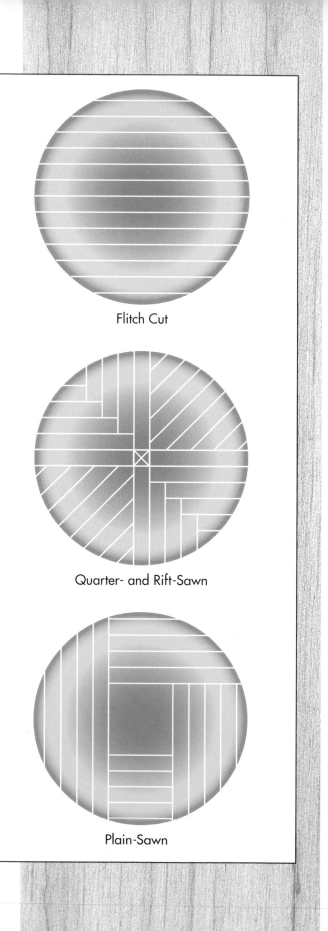

Flitch Cut

Quarter- and Rift-Sawn

Plain-Sawn

As noted in Chapter 1, the orientation of the growth rings in a piece of lumber will determine not only stability, but also board appearance.

Plain-sawn

A plain- or flat-sawn board (*top photo*) is a common type of lumber since this cutting method produces the highest yield from a log. The grain on the face of the board will often swirl in many directions. When a wild-grained piece of wood like this is stained, the softer, more porous earlywood will soak up more stain and be darker than the harder, less porous latewood. The resulting pattern is often referred to as "landscape figure." Flat-sawn lumber tends to move a lot with changes in humidity—it often cups and warps.

Rift-sawn

Rift-sawn lumber (*middle photo*) generally has clearer, straighter grain than plain-sawn lumber. You'll often find rift-sawn lumber in the same stack as plain-sawn. In some cases, you may even see both types of wood in a single board; sawyers call this a "bastard" cut. The face grain will have wilder grain on one side and straighter, more even grain on the other. Generally, it's best to avoid boards like this because the two sides will react differently to changes in humidity. True rift-sawn lumber will be much more stable than plain-sawn, with less tendency to warp.

Quartersawn

Of all the cutting methods, quartersawing produces boards that are the most stable. Quartersawn lumber (*bottom photo*) shows the straightest grain and in some species exhibits ray fleck. Ray fleck is common in quartersawn white oak, red oak, sycamore, and cherry. But beauty like this exacts a price. Quartersawing is the most wasteful way to cut a log, it's time-consuming and expensive, and you need a larger log to produce reasonably wide boards.

HARDWOOD DIMENSIONS

Hardwood lumber is sold by the board foot, a *volume* measurement that indicates thickness and width as well as length. One board foot equals 144 cubic inches—this is easy to visualize as a board 1" thick, 12" wide, and 12" long (*see drawing at right*). This also means that two boards of the same thickness can have different widths and lengths and still have the same board footage.

To determine the number of board feet in a given board, first measure the width and length in inches (it's best to round off fractions). Next consider the thickness. Boards that are 1" or less are all said to be 1" thick. (Note: Boards $1/2$" thick or less are typically sold by the square foot—a surface measure that does not account for thickness.) To calculate board footage, multiply the width times the length times the thickness. Then divide by 144. Say, for example, a board is $3/4$" thick, 8 feet long, and 8" wide. The calculation would be 1" × 96" × 8" =

768". Dividing by 144 gives $5 1/3$ board feet.

Boards thicker than 1" can get confusing. The cause of the confusion is the "quarter" designation system that is still used today but is based on the cutting capabilities of some of the first sawmills. These rather crude water-driven machines could cut only in $1/4$" increments.

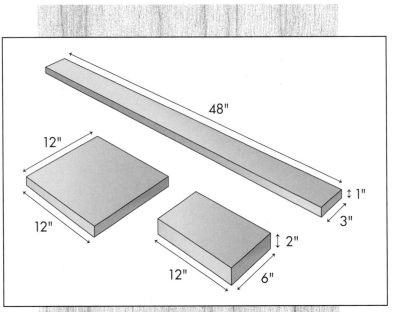

QUARTER DESIGNATIONS

Quarter designator	Rough thickness	Rough thickness	Surfaced thickness	Surfaced thickness
4/4	1"	25.4 mm	$13/16$"	20.6 mm
5/4	$1 1/4$"	31.8 mm	$1 1/6$"	27 mm
6/4	$1 1/2$"	38.1 mm	$1 5/16$"	33.3 mm
8/4	2"	50.8 mm	$1 3/4$"	44.4 mm
10/4	$2 1/2$"	63.5 mm	$2 1/4$"	57.2 mm
12/4	3"	76.2 mm	$2 3/4$"	68.9 mm
14/4	$3 1/2$"	88.9 mm	$3 1/4$"	82.6 mm
16/4	4"	101.6 mm	$3 3/4$"	95.2 mm

Straight-Line Rip

S2S

S4S

Because of this, lumber was sold in "quarters"—three-quarters ($^3/_4$"), four-quarters (4/4 or 1"), six-quarters (6/4 or 1$^1/_2$"), and so on. It's important to note that this measurement is for rough, unsurfaced lumber.

But most hardwood is sold surfaced—this is where the confusion comes from (*see chart on page 62*). Say you want to buy a board that's 6" wide, 8 feet long, and 1$^5/_{16}$" thick. You might think the board footage would be 5$^1/_4$ (6" × 96" × 1.3125"). But it's not—it's 6 board feet (6" × 96" × 1.5"). That's because although the board is only 1$^5/_{16}$" thick, it's 6/4 lumber (it was 1$^1/_2$" thick before planing) and the quarter designator—not actual thickness—is what's used to calculate board feet.

Random widths and lengths

Since most hardwood logs are cut for grade, the lumber that comes off the log can vary in width and length. Because too much good wood would be lost trying to convert each board to a uniform dimension (as in softwood lumber; *see page* 64), hardwood lumber is sold in random widths and lengths (*top photo*). This makes buying hardwood lumber more of a challenge (*see Chapter* 6).

S2S and S4S

Much of the lumber you'll find in lumberyards and at some sawmills will be sold planed or surfaced, instead of rough. The two most common surfacing treatments are S2S (surfaced two sides) and S4S (surfaced four sides); *see drawing at left*. An S2S board has both faces planed. An S4S board has both faces planed and both edges either jointed or straight-line ripped to provide flat, true reference for machining. Be aware that not all mills plane their stock to the NHLA (National Hardwood Lumber Association) guidelines for thickness shown in the chart on page 62. You might encounter 4/4 lumber that's $^3/_4$", $^{25}/_{64}$", or $^{13}/_{16}$". Depending on your project, this may or may not be a problem; it's always a good idea to ask in advance what thickness they plane down to.

SOFTWOOD DIMENSIONS

Because most softwood lumber is used in construction, it is cut to standard sizes. This way architects and builders around the world can all use the same "building blocks" when they design or build a structure. The length of a softwood board is given in actual dimensions, and the width and thickness are given in "nominal" dimensions; actual dimensions are somewhat less. Nominal dimensions are based on rough-cut, green lumber; actual dimensions describe boards after they've been dried and surfaced on all four sides (*see photo on page 65*).

Although softwood lumber is also manufactured in 1-foot increments, 2-foot increments are more common. The width of softwood lumber varies from 2" to 16", usually in 2" increments. Thickness is generally categorized into three groups: *boards* are lumber that's less than 2" in thickness, *dimension lumber* ranges from 2" to 5", and *timbers* are more than 5" (*see the chart below*).

Big mills

Most of the sawmills that cut softwood lumber up into boards, dimension lumber, and timbers are huge. I visited a sawmill in eastern Washington State where the "mill" was around 1/3 mile long (500 meters). Logs went into one end, and finished boards literally came out the other end.

STANDARD SOFTWOOD DIMENSIONS

Item	Nominal thickness	Dressed thickness	Dressed thickness	Nominal width	Dressed width	Dressed width
Boards	1"	3/4"	19 mm	2"	1 1/2"	38 mm
	1 1/4"	1"	25 mm	4"	3 1/2"	89 mm
	1 1/2"	1 1/4"	32 mm	6"	5 1/2"	140 mm
				8"	7 1/2"	184 mm
				10"	9 1/4"	235 mm
				12"	11 1/4"	286 mm
Dimension Lumber	2"	1 1/2"	38 mm	2"	1 1/2"	38 mm
	2 1/2"	2"	51 mm	4"	3 1/2"	89 mm
	3"	2 1/2"	64 mm	6"	5 1/2"	140 mm
	3 1/2"	3"	76 mm	8"	7 1/2"	184 mm
	4"	3 1/2"	89 mm	10"	9 1/4"	235 mm
	4 1/2"	4"	102 mm	12"	11 1/4"	286 mm

A computer on the front end examined each log and then calculated the maximum yield, and off went the log to be cut up into pieces, dried, and surfaced—all in one machine. All that was left was to sort, grade, and palletize the lumber.

By the board foot

Although the majority of softwood is cut and sold in standard dimensions, you'll still occasionally come across a sawmill or lumberyard where they cut a log for grade and sell the lumber by the board foot. In cases like this, the calculation method is the same as that used for hardwood lumber. Measure the width and length and thickness in inches, and multiply them together. Then divide by 144 to get the board footage (*see the equations below*). Here again, you'll need to use the rough or nominal thickness in your calculation.

Surfacing options

Most softwood lumber is surfaced on four sides (S4S). But in the case of some softwood lumber, the faces are left rough and one or two edges are surfaced (S1E or S2E) by either straight-line ripping or jointing. An example of this would be rough cedar boards used for exterior siding and construction.

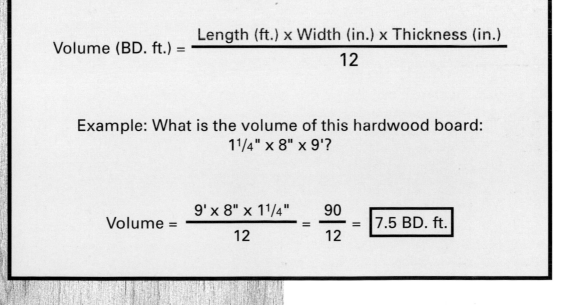

$$\text{Volume (BD. ft.)} = \frac{\text{Length (ft.) x Width (in.) x Thickness (in.)}}{12}$$

Example: What is the volume of this hardwood board:
$1^{1}/_{4}" \times 8" \times 9'$?

$$\text{Volume} = \frac{9' \times 8" \times 1^{1}/_{4}"}{12} = \frac{90}{12} = \boxed{7.5 \text{ BD. ft.}}$$

AT THE SAWMILL

Hardwood sawmills, often called "grade" mills, can be found wherever hardwood grows. These mills vary from two-man portable mills to large mills costing hundreds of thousands of dollars where a team of workers quickly turns logs into lumber.

The yard

Regardless of where a log comes from, its first stop is the yard (*top photo*), where a worker scans it for metal using a hand-held metal detector. Larger mills often have a detector built into the conveyer line. Metal in a log is disaster to a sawmill. When a spinning blade hits metal, the first thing that goes is its teeth. Even worse, there can be severe damage: The blade can distort, bend, or break, resulting in flying shrapnel.

Loading the log

After a log is cleared for metal, it's loaded into the mill with a forklift. Then, depending on the level of mechanization, the log is positioned to be fed into the debarker. At small mills, this is usually done manually using a special tool called a cant hook (*middle photo*). On larger mills, the log is moved via hydraulics.

Debarker

Stripping the bark from a log or "debarking" does two things. First, it gives the sawyer a clear view of the log so he or she can identify defects and position the log for the best cut. Second, it removes dirt, rocks, and other abrasive particles that often get trapped in the bark during the logging operation. This saves considerable wear and tear on the mill's saw blade. Debarking can be done by blasting bark off with high-pressure water jets or by grinding it off with a traveling abrasive-head debarker (*bottom photo*). Removed bark is usually collected and burned to heat the mill or to generate steam for the kiln.

On to the head rig

The now naked and slightly chewed-up logs travel down a conveyer belt to the carriage or "head rig" (*top photo*). Here the logs are clamped securely to the carriage with hydraulic pinchers called "dogs." The carriage moves back and forth and presents the log to the spinning saw blade, where boards are trimmed off. This action is similar to that of the meat slicing machines at most delis.

The sawyer "reads" the log

The sawyer is the most highly skilled member of the sawmill's team. A good sawyer can "read" and judge a log in just seconds. It's the sawyer's job to get the most out of the log. Positioned high above the log, the sawyer uses joystick-like controls to manipulate it (*middle photo*). A flick of the wrist, and the log is rotated. Another flick and the log spins again. The sawyer's controls are linked to hydraulic arms that push and pull the log to virtually any position.

Determining the cut

Once the sawyer decides on the best log position for the first cut, he or she uses built-in lasers to project lines onto the log to show where this cut will be (*bottom photo*). This first cut is critical and will determine how the subsequent cuts are made. Depending on the type of sawmill, laser lines can show both thickness and width if the mill is equipped with multiple blades.

Edging blades

On some sawmills (*like the one in the top photo*), the sawyer can move a set of horizontal blades up and down independently to match the position of the laser lines. This allows the sawyer not only to set the thickness of the board, but also to cut the board to width at the same time. Here again, a flick of the sawyer's wrist is all it takes to hydraulically move the blades in or out and up or down for the cut.

Cutting the log

With the log and trim blades in final position, the sawyer signals for the carriage to push the log into the blade (*middle photo*). Circular-saw blades on mills this size range from 30" to 60" in diameter. The actual cut takes 1 to 2 seconds. At the end of the cut, the boards fall over onto a conveyer belt and the carriage returns to its starting position. Depending on the log, the sawyer may continue cutting boards or may rotate the log to make the next cut.

Circular-saw blades

Although fast and powerful, circular-saw mills do have one disadvantage. Every time a cut is made, roughly $\frac{1}{4}$" of wood is lost to the kerf that the saw blade makes. This may not seem like much, but for every four cuts a sawyer makes, a whole board is lost. This is one area where a bandsaw mill is far superior (*for more on this, see pages 70–73*). The sawing capacity of some sawmills can be increased by vertically stacking saw blades on top of each other (*bottom photo*). This allows them to cut larger-diameter logs without having to make partial cuts.

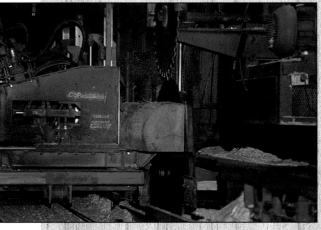

Cutting subsequent boards

The way that subsequent boards are cut off the log will depend on the mill and the equipment that the mill has down the line from the saw. Small commercial mills (*like the one in the top photo*) combine all the cutting and trimming operations in one machine. Typically, the log is cut into a square billet, then sliced into same-sized boards. At larger mills, the sawyer cuts only slabs. These slabs are then sent down the line to an edger, who trims to width, and a trimmer, who cuts the boards to length. The advantage to this type of system is that they can adjust the cut to maximize each individual slab so that there's less waste.

Sorting boards

After the boards come off the mill or down the conveyer from the edger and trimmer, they're sorted and stacked. At some mills, the boards travel down a long conveyer and are graded before going to the kiln. This conveyer is typically referred to as a "green chain" because the wood is wet. Boards may be separated by species, size, grade, or any combination of the three. The boards are stacked on pallets to be moved to the kilns via a forklift (*see Chapter 5 on drying lumber*).

No waste

Every mill I've ever visited generates almost no waste—virtually everything they produce is either sold or reused. Peeled-off bark is chipped and sold as landscaping bedding. Cutoffs drop onto a conveyer (*bottom photo*) and are ground up into chips for bedding or to be burned for heat. Even sawdust is often sold for a variety of uses, including making wood pellets for heating stoves.

A Bandsaw Mill

Unlike their expensive circular-saw cousins, bandsaw mills can be purchased for a fraction of the cost ($20,000 to $40,000, depending on features). Plus, because they're portable, you can take the mill to the log. And a thinner blade ($^1/_{16}$") means a smaller kerf and more lumber.

End-coat logs

The first thing that many bandsaw mill operators do when they receive a log is to end-coat it (*top photo*). That's because many of them cut wood in their spare time, and it can be a while before the log is cut. Without end-coating to seal the logs and reduce checking, the logs could degrade to the point of being useless. Although wax-emulsion coatings designed just for sealing green wood are available, regular latex house paint or even white glue will do in a pinch.

Check for metal

Just as with the big mills, metal in a log is devastating to a bandsaw mill. That's why every mill operator I know checks each log for metal with a metal detector before cutting (*middle photo*). If the metal is near the surface, they'll often make plunge cuts with a chainsaw to remove it. If it's deeper, they may decide to remove that portion of the log. If they find a lot of metal, the log is set aside for firewood.

Make first cut

A smaller mill means less hydraulics, and so the log is often loaded with a forklift combined with muscle power. Once it's on the mill, the operator rotates the log to determine the best cut. Note that the bark is still on the log (*bottom photo*). This makes it difficult to "read" the log, especially for a novice operator. Experienced operators will always take the time to adjust the position of the log so that all cuts will be parallel to the heart (*see page 72 for more on this*).

Cut boards

Just like a larger mill, a bandsaw mill operator will continue rotating a log and cutting until there is a square billet (*top photo*). The big difference here is that after the log is locked in place, it remains stationary and the saw comes to it—the entire bandsaw travels along a track to move the blade through the wood. After every cut the saw returns to a starting position, and the operator adjusts it for the next cut and sends it down along the track again. Another difference is that cut boards don't drop off onto a conveyer. Instead, either they're pulled off manually or an arm on the saw drops down to drag the board back as the saw returns to its starting position.

Flip up to trim

If the boards weren't cut from a square billet, the edges will need to be trimmed straight. To do this, the operator flips one or more boards up on end and clamps them in place (*middle photo*). Even with the aid of hydraulics, sawing lumber with a bandsaw mill like this is labor-intensive. Bandsaw operators and their helpers quickly develop strong backs and muscular arms.

Trim to width

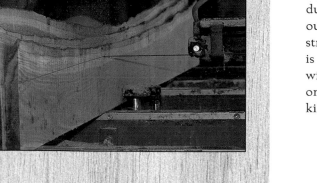

With the boards in place, the operator adjusts the saw for a straight cut that will produce the maximum amount of clear lumber out of each board. After one edge is trimmed straight, the board (or boards) is flipped, the saw is readjusted, and then the board is cut to final width (*bottom photo*). Cut boards are then stacked on a pallet to be stickered before going to the kiln, or set aside for air-drying.

QUARTERSAWING

Bandsaw mills excel at quartersawing lumber for two reasons. First, although quartersawing yields less lumber than the other methods, a thinner blade with its narrow kerf reduces waste and increases yields. Second, bandsaw mills can handle smaller logs and pieces of logs more adeptly than larger mills can.

Parallel cuts are essential

By the very nature of how they grow, all logs are tapered to a certain degree. If the mill operator were to simply roll the log onto the mill and start cutting, none of the cuts would be parallel to the heart. This would result in lumber with angled grain that would be guaranteed to warp. Knowledgeable operators adjust the log to compensate for taper before cutting. To do this, they measure up from the bed to the heart or pith of the log on both ends (*top photo*). Then, using wedges or a built-in toe board, they adjust the log until the difference between the two measurements is less than ¹/₂".

First third

There are two common cutting strategies used for quartersawing. The first cuts the log into equal fourths, and then each triangular-shaped section is resawn. Since gripping these odd-shaped pieces is difficult for the mill, some operators choose a different approach. As shown in the bottom photo, the first cut made removes the top one-third of the log. This is then set aside and will be quartersawn later.

Flitch cut

With the first third removed, the operator then begins cutting the log as if it were

being plain-sawn, or cut through and through. That's because the grain on both sides of the pith will generate the truest quartersawn lumber. The operator will continue cutting through and through until about one-third of the original log remains. This will also get set aside for quartersawing later.

Cut out quartersawn

The next step is to extract the quartersawn lumber out of each of the flitch-cut slabs; these will be the widest and typically the finest boards of the log. Since this wood is sold at a premium, the operator will often cut each slab independently. Each piece is flipped on its edge, clamped in place, and ripped as close to the pith as the grain allows. Then each half is trimmed again to create a straight edge and then is cut to width.

Slice up first third

Each of the log thirds that were cut earlier are now clamped vertically on the saw and boards are cut off in succession (*middle photo*). This produces narrow boards to begin with that gradually get wider the closer the cuts are to the center. After the center is reached, the boards begin to narrow again. Although not as wide as the boards cut from the through-and-through slabs, these boards will likely be quartersawn, or rift-sawn at the very least.

Trim the edges

The final task for quartersawing the log is to cut straight edges on the boards that were cut from the thirds. Here again, the mill operator will flip the boards on edge, clamp them in place, and rip them to final width (*bottom photo*). A good sawyer will make sure that the customer will always get at least one straight edge.

A CHAINSAW MILL

If you don't have $20,000 for a sawmill, how about $200? Inexpensive chainsaw mills attach to a chainsaw and allow you to cut your own lumber—that is, if your back is strong and your chainsaw is powerful enough. There's no doubt about it, chainsaw milling is hard work. A couple of things can help, though: a powerful chainsaw with a 60cc to 70cc engine, and a ripping chain instead of the standard crosscut chain.

Most milling attachments hook up to a chainsaw without your having to modify the chainsaw. Just insert the saw in the mill and secure it with built-in clamps—one near the engine, the other at the tip of the bar. Mills are available in various sizes ranging in capacity from 24" to 56". A 30" mill with at least a 60cc engine is a good starting point. The real expense, though, is the chainsaw. Professional-strength chainsaws (those above 60cc) range between $400 and $1,000. You may be able to get by with a smaller saw, or you might burn it up within the first couple of logs.

The guide rail

The first step in chainsaw milling is to build a guide rail. The guide rail attaches to the log and ensures that the first very important reference cut is flat and straight. It's nothing more than a pair of straight 2×4s held together with aluminum angle or angle iron (*see drawing above right*). Some chainsaw mill manufacturers sell premade aluminum brackets. To provide support for the mill at the beginning and end of the cut, make sure that the 2×4s extend 6" from each end of the log.

Attaching the rail

The guide rail is screwed to the log through the angle brackets (*bottom photo*); use plenty of long screws to create a solid foundation. Before you attach the guide rail to the log, crop off any protruding limbs as close to the log as possible. Then check to be sure you'll be cutting parallel to the heart by measuring from the pith to the bottom of the angle brackets on both ends of the log. Insert wedges as necessary between the guide rail and the log to make these measurements as close as possible.

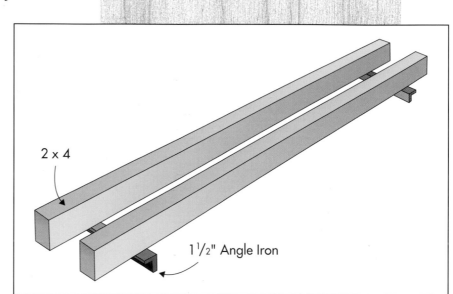

2 x 4

1 1/2" Angle Iron

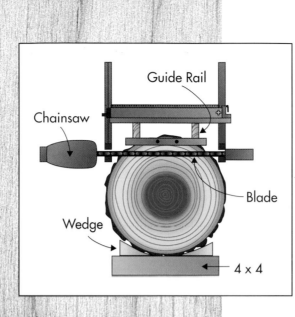

Setup

Before you begin cutting, it's best to elevate one end of the log so that you'll be using gravity to your advantage by milling downhill. Make sure the log is wedged firmly in place and won't rock as you cut. Now adjust the mill for the first cut: Set it to take off an inch or two—all you need to do is create a flat reference (*top drawing*).

The first cut

Now the fun begins. Wearing appropriate safety gear (goggles, ear plugs or muffs, leather gloves, and sturdy boots), start the chainsaw and lift it onto the guide rail. Pull the trigger and gently ease the chain into the log. Keep the mill in firm contact with the guide rail at all times. Don't press to saw too hard; let the chain do the work. Your job is to guide it. The going will be slow, as you're removing a lot of wood (*middle photo*).

The second cut

Depending on how you're cutting the log, you may or may not want to make a deep second cut to create flat parallel sides. If you're cutting through and through, just remove the guide rail, reset the mill for the thickness you want, and keep on cutting. If you want to create a square billet (as I did), remove the guide rail and set the mill for a deep cut (*bottom photo*). As you make this cut, stop periodically and insert wedges in the saw kerf to prevent the weight of the log from pinching the chain as it cuts.

Setup for the third cut

Now rotate the log 90° and use wedges to secure it firmly in place. Place the guide rail back on the log (you'll likely need to move the 2×4s closer together). Use a carpenter's square or framing square to position the guide rail so the cut will be 90° to the face of the sides (*top drawing*). Take your time here and make sure to double-check both ends before securing the guide rail. Here again, it's important to adjust the guide rails with wedges as necessary so you'll cut parallel to the heart.

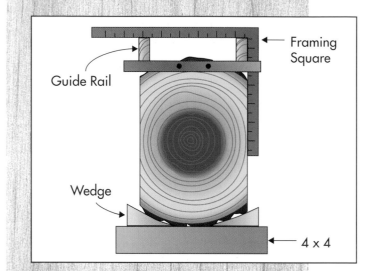

Framing Square

Guide Rail

Wedge

4 x 4

The third cut

Reset the mill to a depth that will cut all the bark off and leave a smooth, flat surface. Here again, ease the saw into the cut and take it easy (*middle photo*). The going will be much better here since you've already removed a considerable amount of wood when you cut the sides parallel.

Cutting boards

With a flat reference, you can remove the guide rail, reset the mill for the desired thickness of boards, and begin cutting. Keep the mill centered on the log, and make sure that it's pressed firmly against the reference surface as you cut (*bottom photo*). After you've cut all the boards you can from the log, you can stack and clamp them on edge and trim them to any width you want.

SMALL LOGS INTO LUMBER IN THE SHOP

If you've always wanted to mill your own lumber but the thought of a howling chainsaw makes you cringe, consider milling a small log in your shop. All it takes is a bandsaw, a router, and a simple shop-made shooting box. The basic idea here is to split a log into manageable pieces, create a flat surface for reference with a router and a shooting box, and then cut this flattened piece into boards on the bandsaw. The size of the log you can handle will depend on the cutting capacity of your bandsaw; this will also determine how large a shooting box you'll need to build.

Shooting box

The shooting box holds the log securely so that you can rout a flat section (*see drawing*). The shooting box here is just a suggestion; you can make it any size you want—just keep the width double the depth. The shooting box I built can handle a 12" diameter log that's up to 30" long. Cleats on the end of the box help prevent it from racking and also provide a convenient place for securing one or both ends of the log. Additional support is provided by one or two L-shaped support blocks. Each is nothing more than a piece of scrap plywood screwed to a 2×6.

A platform that spans the sides sits on top of the shooting box and accepts your router. In use, a straight bit is mounted in the router and the platform is slid back and forth, and from side to side, to rout a flat surface. Cleats on the ends of the platform keep it from sliding off.

Splitting a log

Before you can use the shooting box, you'll need to split the log to fit the box. This is best done with a wedge and a sledgehammer (*bottom photo*). If possible, use a crack or check in the log as a starting point. A few blows with the hammer and you'll be surprised how easily the log splits.

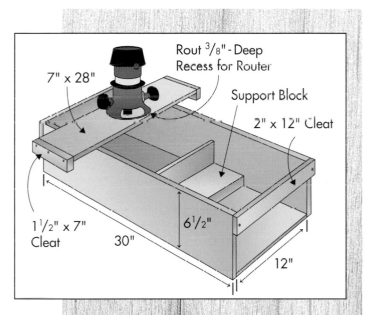

Rout $^3/_8$" - Deep Recess for Router

7" x 28"

Support Block

2" x 12" Cleat

$1^1/_2$" x 7" Cleat

30"

$6^1/_2$"

12"

Secure log to box

Once the log is split in half, set it in the box and secure it. Either screw through the cleats on the end of the box (*as shown in the top photo*) or use the support blocks. With support blocks, you'll need to first screw them to the log and then fasten them to the bottom of the shooting box. The key here is that the log must be close to the top so the router bit can flatten it, but it can't protrude above the sides. If it does, you won't be able to slide the platform back and forth. If the log protrudes and you have a drawknife handy, you can quickly remove the excess. The other option is to remove the log and split off the excess with the wedge and hammer.

Rout a flat surface

With the router mounted on the platform, insert the bit (I used a 3/4" straight bit) and adjust it to take an 1/8"-deep cut. Then turn it on and, starting at one end, slide it from side to side, slowly working your way toward the other end. When the cut is complete, lower the bit another 1/8" and repeat. Continue this way until the entire surface is flat (*middle photo*). This takes less time than you'd think, but it does make a heck of a mess.

Setting up the bandsaw

Now the log is ready for the bandsaw. But first, you'll need to set it up. Two things are critical here. First, you'll need a sturdy rip fence to guide the cuts (*see page 172 to make the fence shown in the bottom photo*). Second, you'll need to install a resaw blade and tension it properly. I used a 1/2"-wide blade with 3 teeth per inch. To prevent bowing, you'll need to apply more tension than you normally would. If you notice the blade bowing as you cut, stop and increase the tension.

Cutting the log

To make the first cut, temporarily attach a piece of plywood to the flat face of the log so the plywood protrudes past the side of the log. By butting this up against the bandsaw's rip fence and cutting, you'll establish a straight edge for the subsequent cuts. Remove the plywood and adjust the rip fence for the board thickness you want. Turn on the saw and ease the log into the blade (*top photo*). Resawing like this takes a firm hand and patience. Go slow and don't force the cut.

Ripping to width

After the log has been cut into boards, you can lower the blade guard, reset the rip fence, and rip each piece to width (*middle photo*). Depending on the log and your project needs, you may want to rip each piece to a different width to maximize the wood in each piece, or you may want to rip them all to the same width.

Stickering

Fresh-cut green wood needs immediate attention to prevent it from checking and twisting. This involves stickering the wood (*bottom photo*) and sealing the ends (*inset*). Set the boards on stickers (kiln-dried wood strips roughly 3/4" square) to help the drying process by allowing air to circulate freely. Coat the ends with green wood sealer or latex house paint (*for more on stickering and sealing wood, see pages* 110–111).

CUTTING BOWL BLANKS

If you like to turn wood and you're tired of paying a premium for the turning blanks, why not try cutting your own blanks? All you need is a small chainsaw and access to downed trees or even stumps. Sources for wood include storm-damage sites, areas being cleared for new homes, firewood companies, nurseries, and tree surgeons. In many cases, the wood will be free or can be had for just a few dollars.

Regardless of the source, the first thing to do is crosscut the log into manageable pieces. A wet log is extremely heavy: A 4-foot section of a 14" to 16" log can easily weigh over 200 pounds. A good idea is to cut the section an inch or two over the maximum size that your lathe can handle.

Draw profile on log

Once the log is cut into smaller sections, the next step is to draw bowl profiles on it with a lumber crayon or a piece of chalk (*top photo*). If you've been turning awhile, you'll know how you want the grain to run. If you're a beginner, just experiment and have fun. This is a great way to learn how grain can be used as an accent.

Cut in half

With the profiles drawn, use a chainsaw to cut the log in half lengthwise (*bottom photo*). Make sure to raise the log off the ground to keep the blade from hitting the ground and causing damage. Wear appropriate safety gear (gloves, goggles, ear protection), brace the log well, and hold the saw at around a 30° angle for optimum control.

Cut to desired width

Now you can cut the half sections of the log as necessary to obtain the desired bowl width (*top photo*). Here again, prop the piece off the ground and hold the saw at around 30°.

Trim to rough shape

After you get the log cut into bowl-sized lengths, you can use the chainsaw to remove excess wood (*middle photo*). Use the bowl profile drawn on the end as a guide to trim away corners, leaving a rough bowl shape. Then if you're working with a long strip (*as shown here*), crosscut the piece into bowl-sized pieces.

Seal the blank

All that's left is to seal the blank. Brush green wood sealer or latex house paint only on the end grain (*bottom photo*). Sealing the other surfaces won't help prevent checking—it'll only slow the drying process needlessly. Store fresh-cut blanks outside, protected from direct sunlight and rain. After six months to a year (depending on the size of the blank), you can bring it into the shop for further drying. If you want to speed this up, rough-turn the blank with 1"-thick walls, seal it completely, and it'll be ready to finish-turn within a month or so. (*See page 111 for another option.*)

"In general, a softwood board is graded as a whole piece, whereas a hardwood board is graded on its useable content aside from any parts which may be considered undesirable for its use such as knots, wane, splits, etc."

NATIONAL HARDWOOD LUMBER ASSOCIATION (NHLA) (1998)

GRADING LUMBER

The idea of "usable content" is something that comes through loud and clear when you take an NHLA lumber grading course, as I did a number of years ago.

"Whaddaya got?" queried the NHLA's chief inspector, Bob Sabistina. "Well, I've got a nice board here that would grade FAS except for those two big knots," answered the novice student grader. "No, no, no," replied Bob, "a good grader doesn't see defects. All he sees is clear wood. Look at the huge clear cuttings you can get out of this board. It's easily No. 1 common."

Like the puzzled student, many woodworkers can't see past the knots to the clear wood a board offers. The day you realize that there is no such thing as a perfect board, free from defects, will be the day your lumber buying gets a lot easier.

As Sabistina advised, don't get caught up in the "defects"—look for the good wood. As the NHLA grading expert says, "An FAS board grades itself." Besides, wood with "defects" sells for a lot less than perfectly clear boards. With just a little bit of extra effort, you can pull the good wood out of a board.

HARDWOOD GRADING

Lumber grading allows anyone who cuts, dries, grades, sells, buys, or uses hardwood lumber to compare apples with apples. Without a grading system, you'd never know what you were buying. What one mill considers "select" might be "common" at another. Fortunately for us, the National Hardwood Lumber Association (NHLA) established a set of guidelines back in 1908 that provided uniform rules for inspecting and grading lumber. To get your own copy of "Rules for the Measurement and Inspection of Hardwood and Cypress," contact the NHLA at www.natlhardwood.org.

Although grading lumber can be complex, the underlying principles are easy to understand. The foundation of the hardwood grading system is the cutting-unit method, which closely approximates the actual manufacturing process where boards are crosscut and ripped into smaller pieces. This method allows a grader to

GRADING RULES

▓ Graders use lumber rules (sometimes called board rules or grading sticks) to determine quickly the total square footage or "surface measure" (SM) of a board. The shafts of these rules are made of thin, flexible hickory so the grader can "read" a board without bending over far. The business end of the rule has a hardened-steel football-shaped disk attached to the end. The edges of the disk are beveled and sharp so the grader can scrape a board to see if a stain is only surface deep. Prebent sharpened points on the end of the disk allow the grader to hook into a board and flip it to determine the best face. (Some rules also have a set of notches in the head to measure a board's thickness.)

Although available in many styles, most lumber rules are 3 feet long and have either 6 or 8 scales running lengthwise on the ruler. Even-length lumber is measured on one side; odd lengths on the other. In use, the head of the rule is hooked over

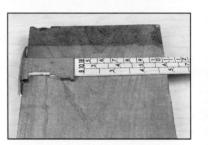

one edge and the rule is pressed flat against the board (*see photo above*). Then the scale corresponding to the length of the board is read to determine the surface measure. If the board shown here is 16 feet long, its surface measure would be 8.

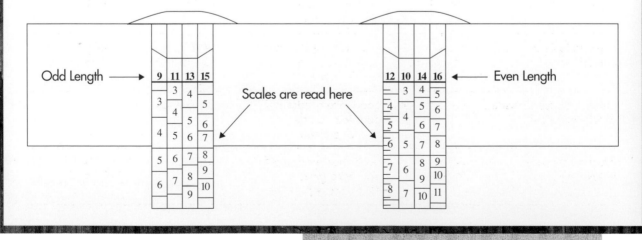

quickly make a mathematical comparison between the total amount of wood in a board and the amount of usable wood the board will yield. The beauty of this is that instead of the grade depending on an arbitrary judgment of the board's appearance, it relies on a mathematical computation that seldom leaves room for interpretation. According to NHLA rules, a cutting is "a portion of a board or plank obtained from crosscutting or ripping or by both. Diagonal cuttings are not permitted."

Basically, the grade of a board depends on the total area of clear cutting the board will yield in relation to its total square footage or surface measure (SM). As a woodworker, you're likely to come across four common NHLA grades: FAS (firsts and seconds), select, No. 1 common, and No. 2 common (actually 2A and 2B); *see the chart below.* Each grade specifies how much clear wood the board will yield: roughly 83% for FAS and select, around 67% for No. 1 common, and 50% for No. 2 common.

With all of the grades except select, boards are judged from their poorer side. A select board is judged from its good side. The select grade is sort of a hybrid of FAS and No. 1 common—the good side must grade FAS, the poor side must grade No. 1 common. Each grade also has requirements for minimum board size, minimum size cuttings, and allowable number of cuts. One important thing to realize about grading is that the clear cuttings in a No. 2 common board are the same quality as those in an FAS board—they're just smaller.

HARDWOOD LUMBER GRADES

	FAS	Select	No. 1 common	No. 2A & 3A common
Minimum size board	6" × 8 ft.	4" × 6 ft.	3" × 4 ft.	3" × 4 ft.
Minimum size cutting	4" × 5 ft. 3" × 7 ft.	good face grades FAS, poor face grade No. 1 common	4" × 2 ft. 3" × 3 ft.	3" × 2 ft.
Basic yield	SM × 10 83⅓%	good face grades FAS, poor face grade No. 1 common	SM × 8 66⅔%	SM × 6 50%
Formula to determine number of cuts	SM÷4 (4 maximum)	good face grades FAS, poor face grade No. 1 common	SM+1÷3 (5 maximum)	SM÷2 (7 maximum)
Surface measure needed to take extra cut	6–15 ft. SM	good face grades FAS, poor face grade No. 1 common	3–10 ft. SM	2–7 ft. SM
Extra yield needed to take extra cut	SM × 11 91⅔%	good face grades FAS, poor face grade No. 1 common	SM × 9 75%	SM × 8 66⅔%

HOW A GRADER GRADES LUMBER

It's amazing how quickly good graders work. In just seconds, they'll read the surface measure (SM), flip the board to determine the best face, make a quick mental calculation, and then assign it a grade. Let's slow this down and examine each of the steps in the process.

Surface measure

The first thing a grader will do is use a lumber rule to determine a board's surface measure (*top photo*). For the four examples shown below, this would be a surface measure of 6 and 8 (*left page*) and 5 and 4 (*right page*). You can calculate surface measure without a lumber rule—just multiply length in feet (rounded to the nearest foot) times the width in inches (and fractions of an inch) and divide by 12.

Determine better face

The next step for a grader is to determine the poor side of the board. This is the side that has the least amount of clear wood. A flip or two of the board is all it usually takes (*bottom photo*). Grading the poorer face helps ensure that the board will meet the minimum requirements for that grade.

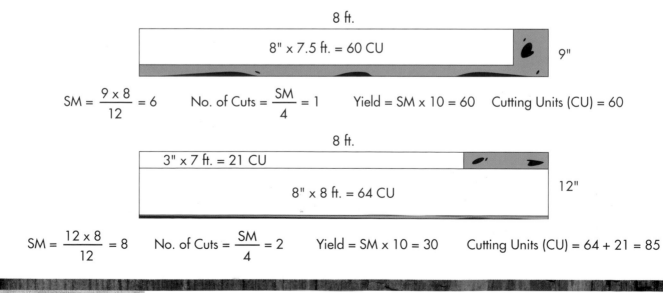

8 ft.

8" x 7.5 ft. = 60 CU   9"

$$SM = \frac{9 \times 8}{12} = 6 \qquad No.\ of\ Cuts = \frac{SM}{4} = 1 \qquad Yield = SM \times 10 = 60 \qquad Cutting\ Units\ (CU) = 60$$

8 ft.

3" x 7 ft. = 21 CU

8" x 8 ft. = 64 CU   12"

$$SM = \frac{12 \times 8}{12} = 8 \qquad No.\ of\ Cuts = \frac{SM}{4} = 2 \qquad Yield = SM \times 10 = 30 \qquad Cutting\ Units\ (CU) = 64 + 21 = 85$$

Calculate cutting units

Here's where experience comes into play. At this point the grader will assign a trial grade to the board, based on an estimate of how much clear wood the board will yield. The grader then identifies clear cuttings and calculates the total yield for the board. In the top example on the left page, one cutting yields 60 cutting units (CU). In the bottom example, two separate cuts will yield 21 and 64 for a total of 85 cutting units.

Determine and mark grade

For the top example on the left page to grade FAS, it needs a minimum of 60 cutting units. At exactly 60, it just makes the grade. In the bottom example, a total of 80 cutting units are needed for FAS. With 85, this is no problem. In either case, if the board doesn't yield sufficient cutting units, the grader drops to the next lower grade and tries again. (*See the boards below for examples of No. 1 and No. 2a grades.*) After a board is graded, the grader marks the board to indicate the grade (*top photo*) and logs it in the tally book (*bottom photo*).

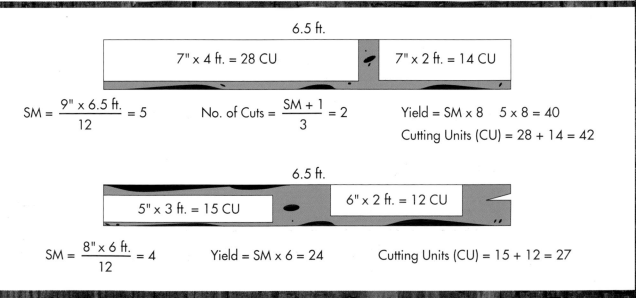

6.5 ft.

| 7" x 4 ft. = 28 CU | 7" x 2 ft. = 14 CU |

$$SM = \frac{9" \times 6.5 \text{ ft.}}{12} = 5 \qquad \text{No. of Cuts} = \frac{SM + 1}{3} = 2$$

Yield = SM x 8   5 x 8 = 40
Cutting Units (CU) = 28 + 14 = 42

6.5 ft.

| 5" x 3 ft. = 15 CU | 6" x 2 ft. = 12 CU |

$$SM = \frac{8" \times 6 \text{ ft.}}{12} = 4 \qquad \text{Yield} = SM \times 6 = 24 \qquad \text{Cutting Units (CU)} = 15 + 12 = 27$$

Hardwood lumber grades specify the minimum requirements that a board must meet to make the grade. Unless otherwise noted, grading is determined from the poor side of the piece.

FAS boards

The highest grade assigned to hardwood lumber, FAS boards yield a minimum of around 83% clear cuttings (*top photo*). With a minimum board size of 6" × 8 ft., and cuttings that must be at least 4" × 5 ft. or 3" × 7 ft., it's no wonder that most FAS boards virtually grade themselves. Take the minimum size board as an example. A 6"-wide, 8-ft.-long board only allows one cutting. This cutting must produce 40 cutting units (CU), which means it can either have a 5"-wide by 8-ft.-long cutting (40 CU) or a 6"-wide by 7-ft.-long cutting (42 CU). Not much room for interpretation here.

As it's a premium grade, FAS has the most stringent limitations regarding defects. The pith in inches cannot exceed the surface measure in feet. Wane is limited to less than one-half the board's length. The length of splits in inches allowed must be less than twice the surface measure except when 1 foot or shorter. The diameter of knots or holes (in inches) must be less than one-third the board's surface measure.

Select boards

The next lower grade, select, can be difficult to find at many lumberyards or mills. That's because grading takes longer, as the board is virtually graded twice. The good face must meet FAS standards and the poor side must grade No. 1 common (*bottom photo*). Since time is money, many graders simply assign the board the lower No. 1 common grade and move on. The yield of select is the same as FAS—83%—except the yield is from the good side. And although the size and number of cuttings are the same as in FAS, the boards may be smaller (4" × 6 ft.).

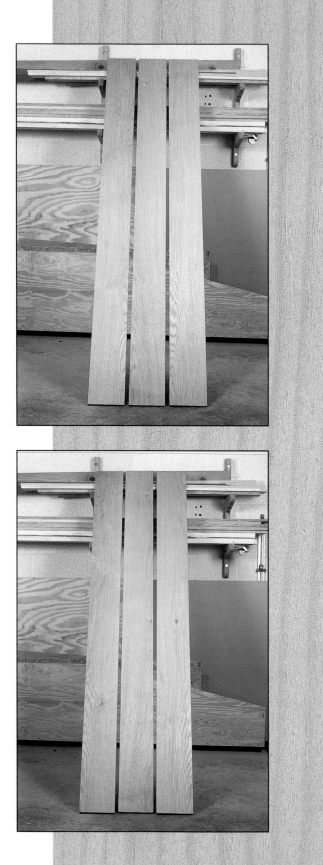

No. 1 common boards

As soon as you drop down into the common grades, you'll notice a lot more defects on the boards (*top photo*)—which means less usable wood (roughly 67% for No. 1 common). Board and cutting size drop dramatically. The minimum size board allowed is 3" wide and 4 ft. long. Cuttings can be as small as 4" × 2 ft. or 3" × 3 ft. Along with this, the number of cuttings allowed increases. If you don't mind a little extra crosscutting and ripping, No. 1 common boards can yield a lot of nice lumber for around half the cost of FAS.

No. 2 common boards

No. 2 common boards are really a combination of two grades—2A and 2B (*bottom photo*). No. 2A common requires clear cuttings, and No. 2B requires only that the cutting be sound—i.e., a cutting that's free from rot, pith, shake, and wane. Sound knots, bird pecks, stain, and streaks are allowed. The yield of No. 2 common lumber is 50%. The minimum size board is the same as No. 1 common, but the minimum size cutting drops to 3" × 2 ft. In order to get the smaller cuttings, more cuts are allowed.

No. 3 common boards

The yield is so low on No. 3 common boards (33%) that they're not much use to a woodworker (or to the mill or lumberyard, for that matter). A lot of work is required to harvest a small amount of wood. Because this grade of lumber is so inexpensive, you'll rarely see it at a lumberyard—there's just no money in it for them.

Softwood Grading

Although you might think that softwood grading would be fairly simple, it isn't. In fact, it's much more complicated than hardwood grading. The main reason it's so complex has to do with its end use. Since the vast majority of softwood lumber sold is used in the construction industry, softwood grading must take into account strength, stiffness, and other mechanical properties of softwood. The problem is, no two woods have identical characteristics. This means every softwood species has its own set of guidelines.

To add to the confusion, there are almost a dozen organizations that publish grading guidelines—most are area-specific or end use–specific. Four of the largest are: National Lumber Grades Authority; Northeastern Lumber Manufacturers Association, Inc.; Northern Softwood Lumber Bureau; and the Western Wood Products Association. Because each grading organization has different names and specifications for the softwood grades, it's almost

impossible to compare apples with apples as you can with hardwood lumber.

Still, there are a couple of things you can do to help unravel the grading mess. First, familiarize yourself with a typical grading system like the Western Wood Products Association's guidelines in this chapter. Second, learn how to read a softwood grade stamp (*see the sidebar below*). Note: Although grade-stamping is often not done on appearance grades of lumber, you may find grade information stapled to the end of a board.

READING A SOFTWOOD GRADE STAMP

■ Most grade stamps have five elements to provide information about the piece. These are: the grading certification, the manufacturer, the grade, the species mark, and the moisture content. Here's how to read the grade stamp shown here.

The circled WWP symbol means that this board has been

graded under the supervision of the Western Wood Products Association. The numeral 12 in the upper left-hand corner is the manufacturer's mill number. The large, bold 3 COM tells us that this board has been graded No. 3 common. The back-to-back P's in the lower right-hand corner are the species mark for ponderosa pine. And S-DRY indicates that the board is surfaced (planed) dry—dried to 19% or less moisture content.

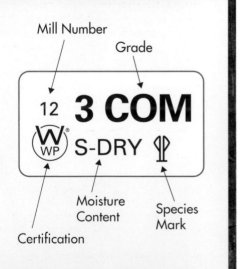

The actual grade for a piece of softwood lumber is calculated with a complex formula that considers the type, size, closeness, frequency, and location of all characteristics and imperfections of the piece. It is the responsibility of the grader to judge visually the total effect of the various combinations according to the limitations in the grading rules for each grade and species.

The softwood grades shown in the chart below are typical of a grading organization and are set forth by the Western Wood Products Association (WWPA). Many of the other grading organizations will have similar names for the grades, but because the species are different, the appearance and characteristics of the grades can vary considerably.

So how do you know what you're buying? Unless you're familiar with the grade stamps and guidelines of the various organizations, you won't know. Here's where dealing with a reputable yard makes all the difference. Although their wood may cost a bit more than at large discount centers, they'll most likely be able to tell you where the wood came from and provide you with detailed grading information. For the most part, the clearer the board, the higher the grade and the more it will cost. (*See pages 92–95 for more information on the various grades*).

SOFTWOOD GRADES FOR BOARDS

Classification	Grade	Appearance
Select	B&BTR	Many pieces are absolutely clear and free from knots; only minor defects and small blemishes are permitted.
	C select	Small defects and blemishes allowed. Recommended for all finishing uses where fine appearance is essential.
	D select	Defects and blemishes are more pronounced; used when finishing needs are less exacting.
Finish	Superior	Only minor defects and blemishes allowed.
	Prime	Similar to superior but with more defects and blemishes allowed.
	E	Pieces where crosscutting or ripping will produce superior or prime grades.
Common	#1 common	The ultimate in fine appearance in a knotty material; all knots must be small and sound.
	#2 common	Contains larger, coarser defects and blemishes; often used for knotty pine paneling.
	#3 common	Broken knots, stain, and knot holes are allowed; used for shelving, paneling, siding as well as fences, boxes, crating and sheathing.
	#4 common	Edge knots, wane, larger knotholes allowed; used widely in construction for subfloors, roof and wall sheathing, and concrete forms.

SOFTWOOD CHARACTERISTICS

Knots play a big role in grading softwood lumber. The WWPA defines a knot as "a portion of a branch or limb that has become incorporated in a piece of lumber." In softwood lumber grading, knots are classified by form, size, quality, and occurrence. A "red" or tight knot is one that results from a live branch growth in the tree that has intergrown with the surrounding wood. A "black" or loose knot is one that results when a dead branch is surrounded by the wood growth of the tree.

Knots are one of the main reasons softwood grading is so complex. A grader must be thoroughly familiar with almost two dozen different types of knots (*see the chart below*). The type, size, number, and location of knots all come into play when a piece is graded.

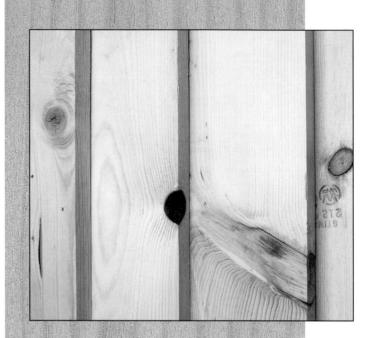

COMMON KNOTS FOUND IN SOFTWOOD

Type of Knot	Description
Round	Produced when the limb is cut at a right angle to its axis
Oval	Produced when a limb is cut at slightly more than a right angle
Spike	Produced when the limb is cut lengthwise or diagonally
Pin or small	Not over ½" or ¾" in diameter, respectively
Medium or large	Not over 1½", or over 1½", respectively
Sound	Contains no decay
Pith	Sound except it contains a pith hole less than ¼" in diameter
Hollow	A sound knot containing a hole greater than ¼" in diameter
Unsound	Contains decay
Firm	Solid across its face but contains incipient decay
Tight	So fixed by growth, shape, or position that it retains its place in the piece
Loose or "not firmly fixed"	Not held tightly in place by growth, shape, or position

To give you an idea of how complex this really is, let's look at just the knot limitations for a 6"-wide No. 3 common board (yes, the number and size of knots allowed changes with the width). For a 6" board, red knots that are sound and tight must be less than 3" in diameter; unsound knots, loose knots, or knotholes must be less than $1^1/4$" in diameter. In addition to this, black knots may be two-thirds the size of allowable red knots, two maximum in each 12 ft. of length, or equivalent smaller, tight black knots. Fixed knots may be equal to knothole size and are limited to two per 12 ft. of length. Only one maximum-sized knothole is permitted in any one piece, but two equivalent smaller knotholes may be permitted if well spread and if the piece is otherwise of high quality. All this covers just the knot limitations for one size board, of one grade, of one species.

In addition to knots and knotholes, there is an array of natural and man-made defects that are considered when grading softwood lumber. Common man-made imperfections are chipped, torn, raised, or loosened grain, skips in surfacing, undersize, mismatch, wavy dressing, and machine-caused burns, chips, bite marks, or knife marks. These rarely affect the mechanical strength of the piece the way natural defects do.

Wane

Wane is the presence of bark or a lack of wood from any cause on the edge or corner of a piece of lumber (*left board in photo below*). Wane that is away from the ends and extends partially or completely across any face is permitted for 1 foot if it's no more serious than skips in dressing allowed, or across a narrow face if no more damaging than the knothole allowed. Wane is limited to one occurrence in each piece.

Shake

Shake is the lengthwise separation of the wood, which usually occurs between or through the annual growth rings (*middle board in photo*). A small amount of shake is allowed in some grades and is classified as light shake (where separation is less than $^1/32$"), or medium shake (separation is less than $^1/8$"). Shake is further defined as "through shake" where it extends from one face to the other, "pith shake" where the shake extends through the pith, and "ring shake," which occurs between growth rings.

Split

Splits are similar to checks except that the separations of the wood fibers extend completely through a piece, usually at the ends. Splits are classified according to their length. A very short split is equal in length to half the width of the piece. A short split is equal in length to the piece's width and cannot exceed one-sixth the piece's length. Medium splits are equal to twice the width and cannot exceed one-sixth its length. A long split is longer than a medium split.

Speck

Speck is caused by a fungus in the living tree and appears as small white pits or spots (*right board in photo*). No further decay will occur unless the wood is used in wet conditions.

STRUCTURAL LIGHT FRAMING LUMBER

There are four grades of structural light framing lumber: select structural, No. 1, No. 2, and No. 3. All pieces are graded full-length. Knots appearing on narrow faces are limited to the same displacement as knots specified on the edges of wide faces.

Select structural

Select structural is the highest grade in structural light framing and is recommended where appearance is as important as strength and stiffness (*three right boards in photo*). Sound, firm, encased, and pith knots are limited to $7/8$" in diameter and must be tight and well spaced. Unsound or loose knots or holes are limited to up to $3/4$" in diameter, one per every 4 lineal feet. Wane, shake, splits, and stain are also limited in this high-quality lumber.

No. 1

When appearance is still important but is secondary to strength, No. 1 grade is the best choice (*middle three boards in photo*). Knots must be of the same type as in select structural grade but can be larger—up to $1^{1}/_{2}$" in diameter. Unsound or loose knots or holes are limited to up to 1" in diameter, one per every 3 feet. Most other limitations are similar to structural select.

No. 2

No. 2 structural light framing lumber is used for general construction (*left three boards in photo*). Well-spaced knots of any quality are allowable up to 2" in diameter, with one hole up to $1^{1}/_{4}$" in diameter every 2 feet. Edge knots are allowable up to $1^{1}/_{4}$". There are still limitations on wane, shake, splits, and stain, but they're not as stringent as for No. 1. and select structural.

No. 3

When strength is not a factor, No. 3 grade can be used. Since strength is less of an issue, knots can be up to $2^{1}/_{2}$" in diameter of any quality, with one hole up to $1^{3}/_{4}$" in diameter every lineal foot. Edge knots are allowable up to $1^{3}/_{4}$" in diameter. And as you'd suspect, limitations for No. 3 grade are more relaxed than for No. 2 grade.

Finish Lumber

The two most common grades of softwood finish lumber you'll find at lumberyards and home centers will be select and common.

Select: B&BTR, C select, D select

Lumber in the C select grade is recommended for projects where a fine appearance is essential (*top photo*). Because its appearance and usability are so close to that of B&BTR (B grade and better), it is often combined with that grade and sold as a C&BTR (C grade and better) select. C&BTR is used widely by woodworkers who enjoy crafting furniture and other projects from pine. Each board may contain only two small, sound knots. Other limitations are similarly strict for C&BTR.

D select grade lumber will still offer a fine appearance, just less so than C&BTR. As a general rule, one side of the board will have a finish appearance and the other side will show more and larger defects. Up to four small, fixed knots are allowed, and the other limitations are relaxed somewhat.

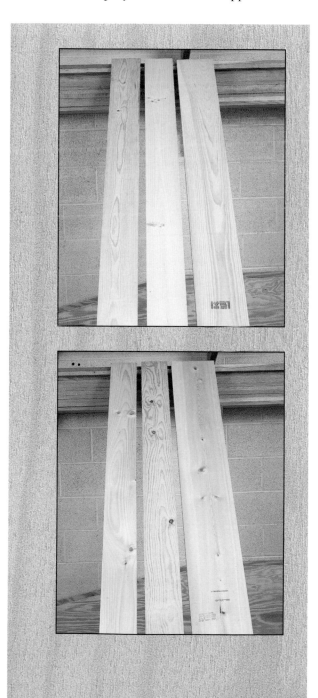

Common: No. 1, 2, and 3

You're not likely to find No. 1 common lumber at most lumberyards. That's because although No. 1 common offers the ultimate in fine appearance of a knotty pine material, it's difficult to keep it stored so that it stays in pristine condition. Most yards will be happy to special-order it for you, but beware, it will be expensive. (You can pay as much for No. 1 common as for some hardwoods.) No. 1 common grade includes all sound, tight-knotted stock, with the size and character of the knot determining its grade.

Although No. 2 common grade is intended primarily for use in housing and light construction, it also works well for country furniture where a knotty pine appearance is desired (*bottom photo*). As long as you seal the knots, this grade takes paint well. Both red and black knots must be sound and tight but can be larger than No. 1 common.

No. 3 common grade is used for a wide range of building purposes where both strength and appearance are important. This grade is often used for shelving, paneling, and siding. It works well for rustic furniture projects, as well. Red knots must be sound and tight and can be larger than in No. 2 common. Unsound knots, loose knots, and knotholes are allowed.

DEFECTS IN WOOD

Warp is defined as any deviation of the face or edge of a board from flatness, or any edge that is not at right angles to the adjacent face or edge. Warp can be either natural or man-made. It occurs naturally because wood doesn't shrink uniformly in all directions. There are considerable differences between the radial, tangential, and longitudinal shrinkage in a piece as it dries (*see pages* 102–104 *for more on this*). Warp is aggravated by irregular or distorted grain and the presence of abnormal types of wood such as juvenile wood and reaction wood (*see pages* 22 *and* 23, *respectively*). Wood can also warp if it hasn't been cut, dried, or stored properly.

The four most common types of warp are bow, cup, twist, and crook (*shown from top to bottom in the drawing*). Bow is an end-to-end curve along the face of a board; it can be caused by irregular grain or improper storage. Cup is an edge-to-edge curve across the face of a board; it often occurs when one face of a board dries more quickly than the other; cup is very common in plain-sawn lumber and on boards cut near the pith. Twist is when one corner of a board is not aligned with the others; it usually results from uneven drying or irregular grain (*bottom photo*). Crook is an end-to-end curve along the edge of a board. It happens because of improper drying or when the pith is near the edge of a board. (*See page* 109 *for strategies for dealing with warped lumber.*)

In terms of hardwood grading, there are strict limitations on how much warp, if any, is allowed. In selects and better, the entire board must be flat enough to surface both sides to standard surfaced thickness. For FAS boards, the entire length, regardless of cutting placement, is considered.

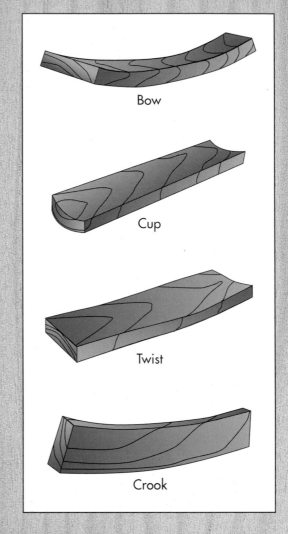

Bow

Cup

Twist

Crook

Stain

Mineral streaks like those on cherry board shown at right in the top photo should not be confused with stain. Stain occurs naturally in many species and is admitted in the cuttings unless otherwise specified in the grading rules. Fungal stains, like the blue stain shown left in the photo, are caused by fungi that grow in the sapwood when the initial stages of drying are too slow. This type of stain is not allowed unless it will dress out in surfacing to standard thickness. Sticker stain is a discoloration that occurs under the stickers in a stack of wood. Sometimes called shadow, these imprints of the stickers can be caused by either chemical or fungal action, or both. In almost all cases, it's considered a defect, since it's unlikely that planing will remove the stain.

Decay

When conditions are right, decay-producing fungi can flourish in wood and attack either the sapwood or the heartwood. Decay is often referred to as rot, dote, or doze and can appear as soft, discolored areas on a board (*middle photo*). Surface growths of decay often appear as fanlike patches. Sometimes fruiting bodies such as mushrooms are produced. In terms of grading hardwood, cuttings must be totally free from rot or decay. Note that this defines cuttings, not boards.

Spalting

Some woods that are attacked by decay develop an attractive dark brown or black staining known as spalting (*bottom photo*). Often referred to as zone lines, these layers can produce amazing patterns when exposed. Although turners love this stuff, it presents quite a challenge to work because the areas between the zone lines will be in varying forms of decay—often soft and punky. The constant change from hard to soft wood and back again makes it tough to cut cleanly. Even the best turners resort to heavy sanding.

FIGURE

Figure is the pattern on a wood's surface resulting from the combination of its natural features and the way the log was cut. The grain direction in a tree is more or less straight, but interesting figure results when the grain is distorted. When a log is cut in a certain way, this figure can be exploited. Surprisingly, the figure a piece of wood displays has little to do with the grade assigned. A board graded FAS that shows stripe, bird's-eye, or ray fleck is still an FAS board. Savvy mill owners, however, will cull this special wood from the stack so they can sell it at a premium.

Ray fleck or flake

The relatively large rays of some species form a conspicuous figure when the wood is quartersawn. Ray flecks (flake, or silver grain, as it's sometimes called) can be delicate, as in the intricate cross striping in cherry, or pronounced, like the large and lustrous rays of white oak in the top photo.

Bird's-eye

In some woods, the cambium can have indentations, bumps, or bulges that leave behind a characteristic figure (*middle photo*). In softwoods, this is commonly called dimpling. In hardwoods, particularly like the hard maple shown here, localized small swirls in the grain direction create the much sought-after bird's-eye figure. No special cutting is required to display bird's-eye figure.

Fiddleback

Wavy grain results in fiddleback figure, so called because it was commonly used for the backs of violins. When wavy-grain wood is cut radially, like the mahogany shown in the bottom photo, fiddleback occurs. The surface looks like a washerboard, caused when light reflects off the surface where the grain intersects at variable angles.

Burl

Burls are wartlike, irregular growths that can form along the trunks or sometimes limbs of trees (*inset*).

Inside the burl you'll find tight clusters of buds. Each of these buds has a dark pith—a branch that never developed, because of stunted growth. The resulting wood tissue within the burl is extremely disoriented and results in a very attractive figure like the walnut in the top photo. Most burls are small, and so burl wood is usually used for small turnings or accents or is sliced into decorative veneer.

Crotch

Crotchwood is highly figured grain that occurs where a limb joins the trunk. As you can see from the walnut in the middle photo, the grain swirls dramatically where the wood fibers have crowded and twisted together. Any tree can produce crotchwood; the angle formed by the limb and trunk is a good indicator as to whether it will be highly figured or not. The wider the angle, the better the chance. When a cut is made down the center of the crotch, the resulting wood is referred to as feather crotch. Cuts toward the outside produce a swirl crotch. Here again, don't plan on using crotch for big projects—the pieces are small or are sliced into veneer.

Stripe

Woods with interlocked grains that slope in alternate directions produce a ribbon or stripe figure when quartersawn (*bottom photo*). The interlocked grain is the result of repeated cycles of spiral growth varying back and forth from left- to right-hand spirals. The stripe effect is caused by the variation in length of severed vessels at the surface. Stripe is often referred to as roe or roey figure; a combination of wavy grain and interlocked grain produces broken stripe or mottled figure.

> "The drying of woods is not a thing to be attempted unadvisedly or indiscreetly. It demands knowledge, care, experience and constant watching."
>
> GUSTAV STICKLEY (1909)

DRYING LUMBER

Stickley knew that wood dried improperly is trouble, and I found out the hard way early in my woodworking career. I had bought six white oak boards, and discovered too late—the only time you can detect faulty drying—that they were case-hardened. I was ripping a 2" thick piece, and halfway through the cut, the kerf closed up so violently that it stopped the blade on a table saw with an industrial-strength, 3-hp motor.

Dangerous? Potentially, sure. Aggravating, time-wasting, and expensive? Absolutely. I ended up with several fine pieces of firewood, and this illustrates one of the bedeviling traits of improperly dried wood: You can't tell till you start machining it. Start cutting into such wood, and you can encounter warp and twist that endanger your safety, your wallet, and your time.

That's why proper drying is so critical to producing usable wood. And that's also why knowing what to watch for—and watch out for—is so important. That's why mills charge for this service. Still, you can save some money by drying wood yourself, as long as you know what you're doing. We'll examine why wood moves the way it does before and after drying... and see why Stickley was so right.

WATER AND WOOD

The lumber cut from a freshly downed tree contains a surprisingly large amount of water. Its moisture content (*see the sidebar below*) can range anywhere from 60% to over 100% (when the water in the wood weighs more than the dry wood itself). For woodworking, the moisture content needs to be down around 6% to 10%. Why? There are several reasons: Wet wood is extremely heavy; it's susceptible to attack by decay and fungi; and if left to dry in uncontrolled conditions, it will cup, twist, bow, or crook (*for more on this see page 96*).

The process of getting wet wood down to 6% has challenged lumbermen for a long time. Over the years, two basic methods of controlling the drying process have been developed: air-drying and kiln-drying. Each has its own advantages and disadvantages. Before exploring these, it's important to first have a solid understanding of how moisture is stored in wood and how it is removed.

How water is stored

Water is stored in two places in freshly cut lumber: the lumens and the cell walls (*see pages 14–18 for more on cell structure*). Water contained within the cell walls is called bound water, as it's held in place by the walls. As wood dries down to around 30%, only the water stored in the lumens is lost—none leaves the cell walls. Eventually, all the water from the lumens has gone—this is a very important point in the drying process called the "fiber saturation point." In most woods this occurs when a moisture content of 27% to 31% is reached. Why so important? This is the point where wood begins to shrink.

CHECKING AND WARPING TENDENCIES OF HARDWOODS

Low	Moderate	High
Alder	Apple	Boxwood
Basswood	Ash	Beech
Birch, paper	Birch, European	Chestnut
Cherry	Birch, yellow	Oak
Cottonwood	Elm, European	Sycamore, American
Elm, white	Elm, rock	
Poplar	Hickory	
Willow	Holly	
	Pear	
	Sycamore, European	
	Walnut	

WHAT IS MOISTURE CONTENT?

■ The term "moisture content" simply describes the amount of water in a piece of wood. For most wood, this is calculated on a dry basis. Moisture content is expressed as a percentage and is equal to the green weight minus the dry weight times 100, divided by the green weight. The dry weight is the weight of the wood that has been dried to the point that no moisture remains. Expressed another way, moisture content is the weight of the water in the wood expressed as a percentage of the dry weight. For example, in a log where the moisture content is 100%, the water bound in the log weighs as much as the dried wood.

Shrinkage

As moisture content drops below the fiber saturation point, the cell walls begin to release water. As the water exits, the cell walls shrink, they move closer together, and the wood becomes stronger (*see top drawing*). The downside to this is that the average tangential shrinkage is around 8% to 10% (the actual amount will depend on the species). Wood also shrinks radially an average of 4%, and almost nothing in length. Since wood shrinks more the twice as much tangentially as radially, the orientation of the growth rings will have a pronounced impact on how the wood deforms (*see middle drawing*).

How wood dries

Unfortunately, the water in cell walls doesn't leave all the cells at the same time. It leaves cells near the surface first (*see bottom drawing*). The center or core of the wood remains wet well after the surrounding wood is dry. This is one of the biggest challenges in drying wood—getting it to dry uniformly.

If you recall the "wood is like a bundle of straws" analogy (*see page* 14), it's easy to see that partially plugging the ends of the straws will help control how quickly the wood releases moisture. This is accomplished by applying a sealer to the ends of boards. Other methods for controlling moisture release include air-drying (*see page* 107) and kiln-drying (*see page* 112). Every species of wood dries differently. Some dry rapidly and behave nicely. Others tend to split, check, or warp (*see the chart on page* 102).

Equilibrium moisture content

Many woodworkers think that once a piece of wood has been dried to, say, 8%, it'll remain that way forever. That isn't true. Wood is constantly reacting to its environment by absorbing and releasing moisture in response to changes in the relative humidity (*see the chart below*).

Where you live also has a sizable impact (*see the maps at right*). Wood is always trying to get to an equilibrium moisture content (EMC), where no moisture is leaving or entering the wood. Wood furniture attains an average EMC of about 6% when stored in homes; that's why wood is dried down to that level.

In reality, wood never reaches a stable equilibrium because the weather is always changing. As the wood absorbs moisture, it swells; as it gives up moisture, it shrinks. Both of these cause a woodworker problems. Understanding why and how wood moves is half the battle; taking steps to deal with it is the other half. *See pages 105–106 for solutions to common wood-movement problems.*

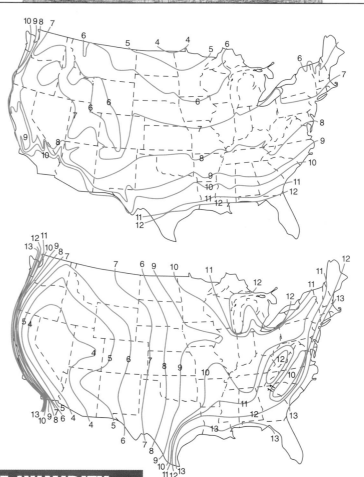

MOISTURE VERSUS RELATIVE HUMIDITY

Relative humidity	Equilibrium moisture content (approximate)
90%	17 – 19%
80%	15 – 16%
70%	12 – 13%
60%	10 – 11%
50%	9 – 10%
40%	7 – 8%
30%	5 – 6%
20%	4 – 5%

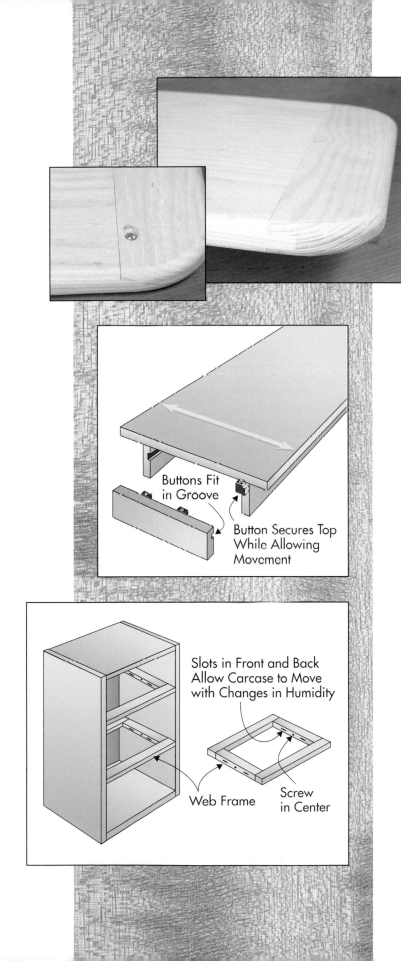

COPING WITH WOOD MOVEMENT

If wood moves constantly, how can pieces be joined so they'll stay together? The answer: Anticipate and compensate for wood movement. We've already covered how it moves (*see pages* 102–104), so let's look at ways to compensate for the movement.

Breadboard ends

The top of a table or breadboard will change considerably in width as it reacts to relative humidity. But since the grain of the end caps is perpendicular to the top, they'll move much less (*top photo*). If you glue the end cap in place, the top won't be able to move; either the top cracks, or the end cap splits. The solution is to glue the cap only at the center and then attach it underneath near the ends with screws set in slotted holes (*inset*).

Buttons

Since a tabletop moves considerably across its width, screwing it directly to a base will surely cause it to crack over time. One way to attach it so it's secure and can move is to use "buttons." Buttons are small, rabbeted blocks of wood that fit into grooves in the table aprons (*red pieces in middle drawing*). As the top changes in width, the buttons slide along with it in the grooves.

Web frames

To lighten a heavy chest of drawers and save money, savvy woodworkers use web frames instead of solid wood for the dividers between drawers (*bottom drawing*). The only disadvantage to this is that the grain of the frame runs perpendicular to the chest sides. Here again, if you glue these in place, the sides can't move without cracking. To prevent this, drill or rout slots near the ends of the frame for screws. When the frame is screwed in place, the sides can move independent of the web frames.

Buttons Fit
in Groove

Button Secures Top
While Allowing
Movement

Slots in Front and Back
Allow Carcase to Move
with Changes in Humidity

Web Frame

Screw
in Center

Multiple tenons

A small but important wood-movement challenge occurs whenever you join an apron to a leg with a mortise-and-tenon joint (such as a table base or bed frame). The problem here is that the tenon grain is perpendicular to the mortise grain inside. If the mortise-and-tenon is small, the movement is insignificant. But with wide tenons, the movement can be enough to break the glue joint. One way to eliminate this is to cut multiple tenons so the movement is spread out over the smaller tenons (*top drawing*).

Floating panels

The frame-and-panel is a staple of cabinetmaking. A grooved frame, typically joined with mortise-and-tenon joints, accepts a thinner panel (often beveled or "raised"). Early cabinetmakers learned that if they glued the panels in the grooves in the frame, the panel would crack. This led to the "floating" panel, where the panel is secured to the frame at only the top and the bottom with a single centered dowel or pin (*middle drawing*). This secures the panel yet allows it to "float."

Stub tenon and groove

A different solution to the frame-and-panel challenge is to use a panel that hardly moves: a plywood panel. Plywood is so stable that you can glue it into the grooves in the frame. Since the glue surface is so large, you don't need to cut mortise-and-tenon joints. Instead, a simpler stub tenon–and-groove joint can be used (*bottom photo*). The beauty of this joint is that after you cut $1/4$" grooves on the inside faces of the frame pieces and $1/4$" tenons on the rails—you're done. The grooves in the stiles also serve as the mortises for the tenons.

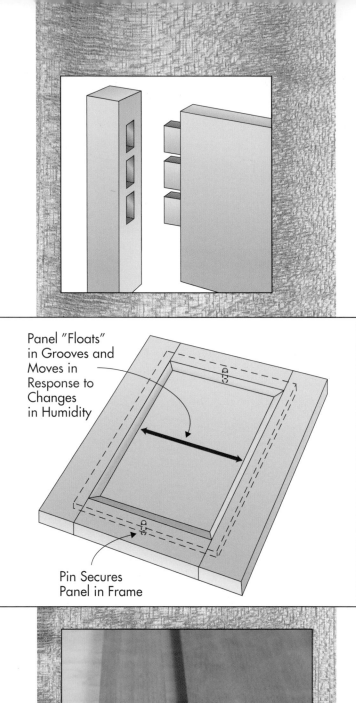

Panel "Floats" in Grooves and Moves in Response to Changes in Humidity

Pin Secures Panel in Frame

AIR-DRYING

Most of the commercial wood sold in the United States is dried in large kilns to speed up the natural drying process (*see pages* 112–114). But even sawmills with kilns often partially air-dry wood before putting it in the kiln. There are two reasons for this: Space is limited in a kiln, and fresh-cut wood will rapidly lose some of its excess moisture naturally if left out in the open (as long as it's stickered, stacked, and protected). The disadvantage to air-drying is time—roughly one year per inch of thickness—plus it takes up yard space. Also, the wood will drop only to a certain percentage outside (roughly 20%). For the moisture content to drop lower, the wood needs to be moved inside.

Foundation

Although some mills simply set a lumber stack on the ground to air-dry, savvy mill owners build stable foundations for their stacks. These will often be cement slabs, old steel rails (*as shown in the top photo*), or leveled timbers. A solid foundation does two things. First, it ensures that the stack remains level. And second, it helps prevent ground moisture from seeping into the stack. The location is just as important. The mill owner selects a spot in the open, where air will circulate evenly around the entire stack.

First layer

Creating a stable, well-ventilated stack takes thought and patience. As a general rule, mill owners use lower-quality wood on the bottom and top of the stack, as these have the greatest tendency to degrade. In most cases, the first layer is 4 feet wide. The mill owner lays down boards, carefully measuring until the desired width is reached (*bottom photo*).

Stickers

When the first layer is down, stickers are added. Stickers are $3/4$"- or $7/8$"-thick strips of dry wood planed to the same size. Starting at one end, a sticker is placed every 16" or so (*top photo*). Many mill owners measure this and mark the outside bottom boards on both sides of the stack. They make sure the stickers are straight and on the marks.

Additional layers

Now the next layer is added. Here's where the challenge to building a stack comes in. The sawmill owner needs to pick boards carefully to end up with the same width as the previous layer. If the mill owner can't find a narrow enough board to fill a layer completely, he or she will keep the outside boards flush with the stack and spread out the inner boards so the gaps are consistent (*middle photo*). For boards that are shorter than the stack, another sticker is added to support the end. If it's left unsupported, it'll surely warp (*see inset*). Stickering continues like this until the stack is complete.

Cover and press

To protect the stack from the elements and help prevent warp, a cover is added to the stack (*bottom photo*). The cover shown here is built from scrap lumber, plywood, and a sheet of corrugated metal. Notice how one end is higher than the other to encourage water runoff. For best protection, the cover needs to extend past the stack at least 6" to 12" in every direction. Now it's just a matter of time.

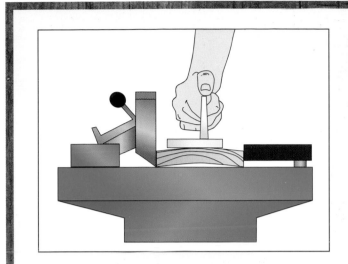

DEALING WITH WARP

No matter how carefully wood is dried, there will always be warped wood in a stack. In cases where the warp is severe, you're best off using the wood to heat the shop. But lumber that has moderate cup, bow, or twist can often be salvaged using one of the three methods described below.

Removing cup

The simplest way to remove cup from a board is to use a jointer (*top drawing*). Start by placing the concave face of the board on the bed of the jointer. This way the edges of the board make solid contact with the bed as you pass the board over the cutterhead. Continue making light passes until the high spots have been removed and the face is flat. With one face flat, you can now resaw or plane the board to the desired thickness.

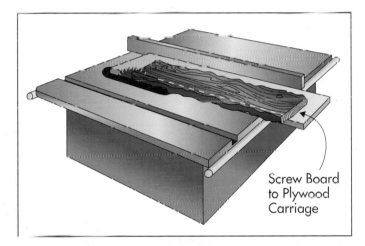

Screw Board
to Plywood
Carriage

Removing bow or taper

Boards that are bowed or tapered can be straightened with a straight-line ripping jig (*middle drawing*). The jig is just a piece of plywood that's wider than the board and at least as long. Screw the warped board to the jig so one edge extends over plywood. Then adjust the rip fence to trim off the overhang. Remove the board, readjust the fence, and trim the board to width.

Removing twist

Twist is the toughest form of warp to deal with. The most effective method I've found is to first crosscut the board into short sections (*bottom drawing*). Then I take these to the jointer and joint one face flat. Keep even pressure on the push block to prevent the piece from rocking. Then place that face against the jointer fence and joint the edge. This creates a 90° edge that can be used to rip and plane the piece square.

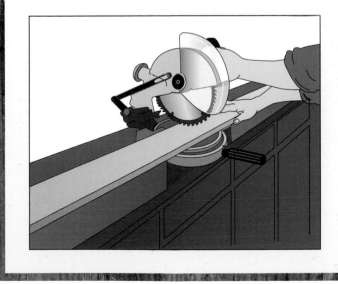

AIR-DRYING YOUR OWN WOOD

Air-drying lumber isn't hard, it just takes a little know-how and a lot of patience. The general rule of thumb is one year for every inch of thickness (*see the chart below for specific times for some common woods*). But different woods dry at different rates, and a moisture meter (*see page 117*) is the best way to track drying speed. In most parts of the United States, air-drying lumber outside will bring the moisture content down to around 20%. At that point, it needs to be moved indoors to continue drying. When it hits 8% to 10%, it's ready to be worked.

Location, location, location

Of all the choices you have to make about air-drying lumber, the location you choose can have the greatest impact. You need a place that will provide good ventilation to the entire stack: You don't want a stack next to a building, since the air flow will be uneven, resulting in irregular

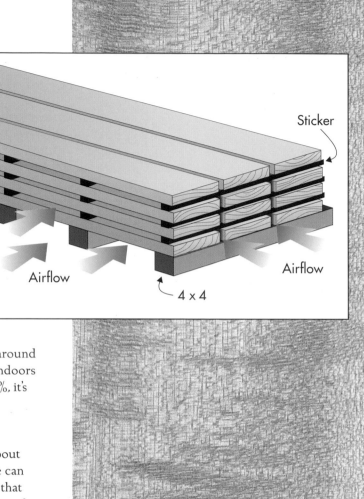

Sticker

Airflow

Airflow

4 x 4

AIR-DRYING TIMES FOR COMMON WOODS

Very slow (9–18 months)	Slow (6–15 months)	Moderate (5–12 months)	Fairly rapid (4–10 months)	Rapid (3–9 months)
Oak, American red	Apple	Birch, yellow	Ash, European	Alder, European
Oak, American white	Cherry, American	Elm, rock	Basswood	Alder, red
Oak, European	Cherry, European	Hickory	Beech, European	Hackberry
	Chestnut	Walnut, black	Birch, European	Poplar
	Holly	Walnut, European	Birch, paper	
	Hornbeam		Cottonwood	
	Pear		Elm, American	
			Elm, European	

drying. A dry foundation is also a must. Use cinder blocks, old steel beams, whatever you have to raise the stack off the ground. Without this, moisture will seep into the lumber, delaying drying and causing mildew, decay, and rot.

The stack

To create the first layer, use lower-quality boards, since this wood tends to degrade. Support the boards every 16", and lay boards down until the desired stack width is reached. There's much debate about how much gap, if any, is necessary between the boards. The industry standard is between $1/2$" and $3/4$". Whatever you use, just try to keep the gaps even.

Sealing the ends

If the ends of the boards haven't been previously sealed, brush on a generous coat of green wood sealer, latex house paint, or white glue (*top photo*). Without sealer, the moisture will evaporate too quickly out of the ends, and the boards will check and crack.

Stickers

Stickers are inserted between layers so air can circulate freely on all surfaces of the wood. There are two basic rules for making your own stickers: The wood must be dry, and all stickers must be the same size. The most common size in industry is $3/4$" square. To help minimize "sticker stain" (light discolorations on the boards from stickers), you can rout a shallow cove on each face of the sticker. This reduces the contact the sticker makes with the drying lumber. Note: Sticker stain is most common on lighter-colored woods, such as maple and ash.

Place stickers every 16" between layers, and align them vertically (*see the drawing on page 110*). Continue stacking until you reach the top layer (again, lower-quality boards are best saved for this). Cover the stack with a sheet of pressure-treated plywood cut to extend past the stack in all directions, and set it at an angle to assist water drainage (*bottom photo*). Check the moisture content of the wood every month or so with a meter. When it hits 20%, move the stack inside to continue drying.

KILN-DRIED LUMBER: DEHUMIDIFICATION

A dehumidification kiln works on the same principle as a dehumidifier in your home. As air circulates through the condenser unit, it cools down and water vapor in the air condenses, collects, and is drained away. In a kiln, the condenser is much larger, the compartment is sealed, and air is forced to circulate. As moisture seeps out of the wood, it is whisked away by the condenser unit. Over a period of weeks, all but a small percentage of the moisture is removed from the wood. When the wood reads 6% to 8% with a moisture meter or probe, it's ready to come out.

Many kilns use probes to monitor the moisture content of the wood. The kiln operator finds a board in the middle of the stack with a defect (not wanting to ruin a good board) and drills a small hole to accept the probe tip (*top photo*). After the probe tip is hammered in place, a cable is hooked to it and fed out the side of the stack (*inset*). Most systems like this allow the operator to monitor multiple probes at the same time. This way, different species can be loaded in the kiln and tracked separately.

With the probes in place, the stack is loaded into the kiln with a forklift (*middle photo*). Then the probe cables are plugged into the monitor. Next, the fans and condenser are turned on (*bottom photo*) and the doors are sealed. Now it's just a matter of time. The kiln operator checks the monitor every day or so and notes the progress in a log. As different species dry, they can be unloaded.

Kiln-Dried Lumber: Steam

Most large kilns in the United States are steam-heated kilns where both temperature and humidity are closely controlled. They use a combination of wet and dry heat to rapidly remove excess moisture from wood. These kilns are used to dry both green lumber and partially air-dried lumber. For hardwoods, the temperatures inside the kiln range from 100° to 180°F; air speeds through the lumber range from 200 to 400 feet per minute.

Although there are high-temperature kilns that use temperatures between 230° and 250°F, they work best with softwoods. When they're used to dry hardwoods, honeycombing, collapse, checking, and darkening of the wood often result. Research shows that a combination of the two may be effective—starting low and then switching to high temperatures.

Big buildings

Steam kilns are really just large buildings where the internal climate can be easily altered and closely monitored (*top photo*). In many cases, the stickered lumber is stacked on pallets and then placed on a large platform that rides on rails set into the floor (*middle photo*). Inside the kiln there are two probes—a "wet bulb" and a "dry bulb." Each controls the amount of wet heat and dry heat, respectively.

Kiln schedules

Although kiln-drying does require a lot of skill, there is help available. The Forest Products Society publishes "Dry Kiln Schedules for Commercial Woods," which contains suggested dry-kiln schedules for over 500 commercial woods. These schedules call for changes to be made to the wet and dry heat on the basis of the average moisture content of the wood. The use of sample boards and accurate record keeping is required (*bottom photo*).

Sample boards

To track how the lumber is drying, the kiln operator periodically measures the moisture content of the wood. This is done with sample boards cut from larger pieces of lumber in the stack; small "moisture sections" are trimmed off of these. These sections are carefully weighed and the moisture content is calculated (*top photo*). Sample boards are chosen to represent all the boards in the stack, from the fastest to the slowest drying, and from the widest to the narrowest. Then the kiln operator checks this moisture content against the kiln schedule. Any changes that need to be made are done at the monitor (*middle photo*) or control panel.

Treatments

Once the lumber reaches the desired moisture content, two treatments may be used on the boards: equalizing and conditioning. Equalizing is used when there's a wide spread in the moisture content between boards (typically 3% or greater). Conditioning is used to relieve the stresses that drying creates within the boards. This stress or tension set (often referred to as case-hardening) occurs because the outer portion of a board dries faster than the inner portion. Conditioning increases moisture in the kiln anywhere from 4 to 72 hours after the desired moisture content is reached.

Dead stack

When either or both treatments are complete, the wood is removed from the kiln, the stickers are removed, and the lumber is "dead" stacked (*bottom photo*). Since the wood is dry, there's no need for air to circulate, and the weight helps keep the lumber flat. Often the species and thickness are marked on the edges of some of the boards in the stack.

BUILDING A SOLAR KILN

If you'd like to dry your own wood but don't want to wait for it to air-dry, a solar kiln may be the answer. A solar kiln uses heat from the sun to accelerate drying. The wood dries during the day, when it's warmed by sunlight. At night, the moisture from the wetter core migrates towards the surfaces. To prevent the boards from overheating on warm days, a thermostatically controlled fan circulates air, and warmer air escapes out a vent.

Choose a sunny location for the kiln, and level the surface. Spread gravel over landscape cloth (to keep out weeds), and use timbers or concrete blocks as a foundation. The solar kiln shown here is a suggested starting point; modify it to suit your needs. After the frame is built, cover the sides and back with exterior plywood. For the top and front, use plexiglass, clear corrugated plastic, or tempered glass (old patio doors work great). Apply caulk to prevent leaks.

A fan is installed on the floor, centered on the back. Install a thermostat centered on the top of the back wall, and adjust it so when the inside temperature exceeds 80°F, the fan turns on. Install vents near the top of the back wall to allow hot air to escape. To keep air circulating under the stack, nail 2×2 cleats to the floor every 16". Stack the lumber as you would for air-drying, and monitor the drying process with a moisture meter.

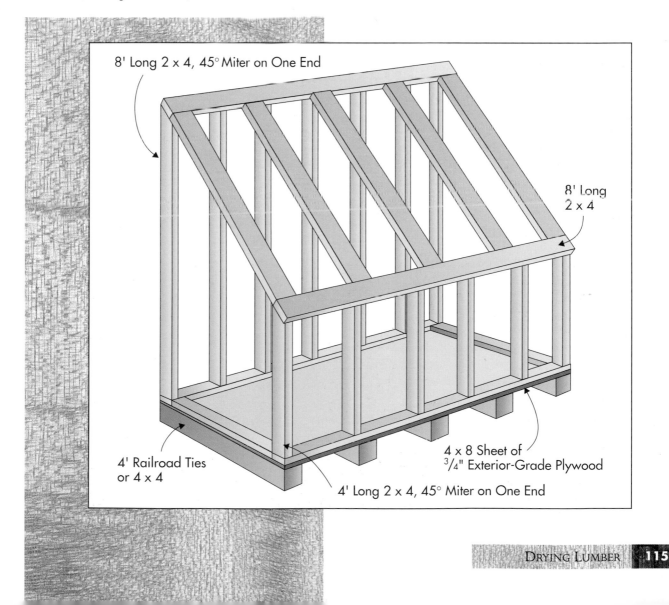

8' Long 2 x 4, 45° Miter on One End

8' Long 2 x 4

4' Railroad Ties or 4 x 4

4 x 8 Sheet of ³/₄" Exterior-Grade Plywood

4' Long 2 x 4, 45° Miter on One End

DRYING DEFECTS

Case-hardening

Case-hardening occurs when wood is kiln-dried. Basically, the surface of the wood dries too rapidly (usually in the early stages), causing permanent set and stresses in the outer zone and tensile stresses in the core. Kiln operators check for this by cutting stress sections, often called a "prong test." Two relief cuts are made into the sample with a bandsaw (*top photo*). If the prongs bow significantly out or in, then case-hardening is present. A good prong test results in straight or nearly straight prongs after 16 to 24 hours. Case-hardening is prevented by the conditioning stage (*see page* 114).

End checks

When the ends of boards dry faster than the rest of the wood, end checks and end splitting result (*middle photo*). This can happen with either air-dried or kiln-dried lumber. The solution in both cases is to apply end sealer as soon as possible. End splitting can occur when there's too much air circulation over the ends and not enough over the stack. This can usually be prevented with proper sticker placement (near the ends) and by using baffles to ensure even airflow.

Surface checks

Surface checking appears in kiln-dried wood when the surface dries too rapidly in relation to the core (*bottom photo*). The best way to prevent this is to use higher relative humidities in the early stages of drying by using accurate temperature- and humidity-monitoring equipment. If surface checking is slight, the checks may close up when the wood is fully dried to a uniform moisture content.

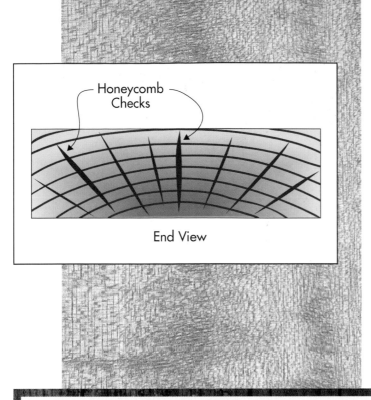

Honeycomb
Checks

End View

Honeycomb

Honeycombing occurs when lumber undergoes severe case-hardening in the early stages of drying. Internal checking results from excessive tensile stresses in the core (*top drawing*). Honeycombing can also be caused by excessively high temperature in the final stages. What's particularly nasty about honeycomb is that it often resembles surface checking, which you might think you could just plane away. If you try this, you'll find that the checks widen the deeper you go. Honeycombing can be prevented by using higher relative humidities in the early stages, applying periodic steaming, and limiting the final temperature. If lumber you've purchased from a mill or lumberyard has honeycomb, don't try to work with it: It's useless. Instead, take it back—it wasn't dried properly.

USING A MOISTURE METER

Moisture meters have been used in the commercial lumber industry for years to monitor the moisture content of wood. Although expensive, they are accurate and reliable, and several manufacturers have made this technology affordable to the average woodworker. There are two basic types of moisture meter available. One is scanned across the board. The other (*as shown in the photo*) uses a pair of short, sharp pins to penetrate the wood. As a rule, meters with pins are less expensive, but they aren't as easy to use.

To use a pinless meter, just hold it over the board and press the test button. For a pin-type meter, push the pins into the surface

until the push-button switch between the pins depresses far enough to turn on the meter and provide a reading. Take your reading away from knots and the ends of boards, which will give you an abnormally high reading. It's also a good idea to take multiple readings and average the results. To achieve accurate results on thicker stock when using a pin-type meter, first drive in nails to "extend" the pins deeper into the wood (*see drawing below*).

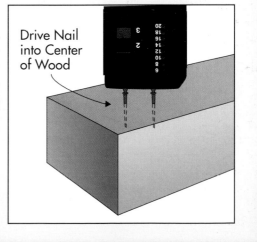

Drive Nail into Center of Wood

> "A good rule is to buy as much as you can sensibly afford of any wood that excites you and then, quickly, buy a little bit more."
>
> JAMES KRENOV (1975)

BUYING and STORING LUMBER

"Where's your rig?" asked the old man at the south Florida lumber-yard. I gestured to my little pickup truck. A young lad in my early 20s, I was stunned. Everywhere I looked, Quonset huts crammed with hardwood lumber awaited their usual customers: construction companies that piled their giant flatbeds with wholesale orders.

I was after a couple of white oak boards to build a tool chest—but I got an education in buying lumber. The old man let me wander around, clambering over the stacks to extract this board and that. "Honeycomb," he'd pronounce, shaking his head. "Good for nothin' but firewood." Or, "Not that board, son; it has sticker stain." He taught me how to use a grading stick, what to look for in wood, how to judge a board.

Now, like every woodworker I know, I hoard the stuff; I'm just more selective than I used to be. I have little bits of exotic woods that I can't throw away. Someday I'll use them.

Let's walk through the stacks with that wise old man. And find out how, when a wood really excites us, to buy it and store it so that our final result is just as exciting.

ESTIMATING LUMBER

In a perfect world, all boards are defect-free, color-matched, and just the right size for all parts of a project. In the real world, hardwood is sold in random widths and lengths, rarely matches in color, and often has knots, checks, wane, and warp. That's why it's critical to have a solid grip on how much and what type of lumber you need for a project before you buy it.

The first step is estimating how much wood you'll need. This is where a cutting list and a cutting diagram can help. If you're working from a set of plans, there will likely be a cutting or materials list. It's important to note that these lists usually show finished dimensions; you'll

need to factor in waste for cutting and planing.

If you've designed the piece yourself, take the time to generate a cutting list. Sometimes it helps to visualize how much wood you'll need by creating a cutting diagram as well (*see the sample shown below*). As a general rule of thumb, FAS boards are around 6" wide. If you plan for 5$^{1}/_{2}$"-wide boards that are 8 feet long, you should be fine. You can use wider boards, but realize that you'll pay a premium for anything over 8" wide, and that wider boards tend to cup.

To make sure you have plenty of wood, buy at least 20% more than you think you'll need. This extra lumber will let you compensate for snipe or mistakes, and allows for the best grain selection (*see the examples on page* 121). And if everything goes well, you'll have a little stash left for another project.

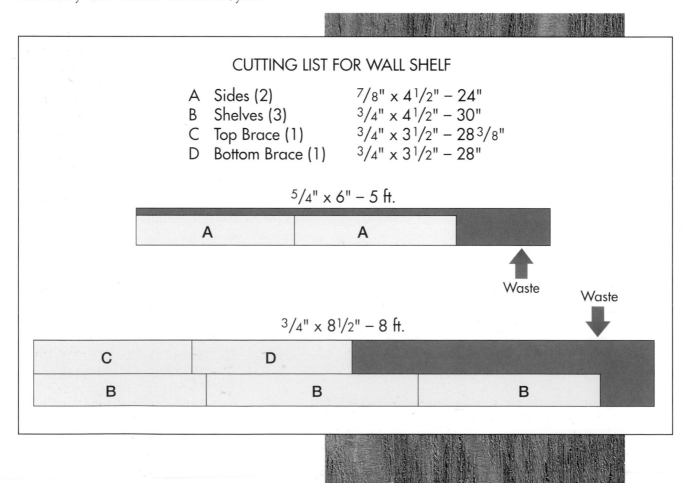

CUTTING LIST FOR WALL SHELF

A Sides (2) $^{7}/_{8}$" x 4$^{1}/_{2}$" – 24"
B Shelves (3) $^{3}/_{4}$" x 4$^{1}/_{2}$" – 30"
C Top Brace (1) $^{3}/_{4}$" x 3$^{1}/_{2}$" – 28$^{3}/_{8}$"
D Bottom Brace (1) $^{3}/_{4}$" x 3$^{1}/_{2}$" – 28"

$^{5}/_{4}$" x 6" – 5 ft.

| A | A |

Waste

Waste

$^{3}/_{4}$" x 8$^{1}/_{2}$" – 8 ft.

| C | D |
| B | B | B |

Snipe

It's said that there's no excuse for snipe—that nasty divot your planer takes out of the last 2 or 3 inches of each board (*top photo*). It can often be prevented by adjusting the feed-roller pressure and supporting the workpiece as it passes into and out of the planer. But even when properly adjusted and with the workpiece supported, many planers still do this. So if you cut off the snipe as you should from both ends, you'll lose 4 to 6 inches of every board. Keep this in mind as you estimate lumber.

Mistakes

Every woodworker makes mistakes. I have a friend who says the difference between a craftsman and an amateur is their ability to hide mistakes. I've even seen talented craftsmen turn a mistake into a design element, like adding an inlay out of contrasting wood to cover a gap. The point is, we all make mistakes like the errant notch cut into the stretcher of the table in the middle photo. Not enough wood caused the builder to patch the stretcher, instead of replacing it. Extra wood could have prevented this.

Best selection

Having extra wood allows you to pick and choose the best wood for a part. It lets you match pieces and use the grain effectively. Notice the care that was taken in selecting the quartersawn white oak for the parts of the Craftsman-style rocker shown in the bottom photo. The ray fleck on both front legs gently curves up; the pattern of ray fleck on the arms gives the impression of bookmatching, although the pieces were cut from different boards. This is possible only when you have extra wood on hand.

SPECIAL PARTS

After you've estimated the bulk of the material you'll need for a project, it's a good idea to identify any parts that will strongly affect the overall appearance of a piece. Care in selection of these highly visual parts can make or break the piece. Consider, for example, any tabletop or the front and side pieces of a cabinet. In the sideboard drawing shown at right, this would entail the top, the side panels, and the fronts of both the drawers and doors.

If you were to pick only enough boards to glue up a top, chances are you'd later wish you had purchased more. I try to set aside at least one or two extra boards in case one of them acts up during machining or I discover a blemish or sticker mark during planing. In the example shown, I would select a board that is wide enough and long enough for all the narrow drawer fronts. This way the grain could run continuously across the face of the piece. At the same time, I would select an alternate in case of problems or mistakes, like miscutting the center drawer front. If you don't have an alternate board, you'll be stuck trying to match the color and grain to the end pieces.

Since you'll usually bring home a stack of wood, it's smart to label the special boards at the lumberyard with a lumber crayon or piece of chalk. On parts that are to be glued up, I'll often go to the trouble to draw the standard cabinetmaker's triangle on them as a reminder of the intended assembly sequence.

I've been privileged to experience some extraordinary woodworking firsthand, from James Krenov's masterful cabinets to Sam Maloof's sculpted chairs. I've judged pieces for design books, art shows, and competitions. Many of the pieces were flawless.

For me, the difference between an exquisite piece and an ordinary one is often how the craftsman uses grain and color of the wood. Every piece of wood is different, and a woodworker can use these subtle differences to advantage.

The half-round table shown in the top photo is a simple example. Notice how the grain used for the sides is continuous and also curves gently to match the curve of the front. This wasn't an accident—it was planned. Attention to details like this makes the difference.

GRAIN AS A DESIGN ELEMENT

Too often, a woodworker will randomly grab a piece of lumber, cut it up for a project, and start gluing pieces together—with virtually no thought to the grain. The grain of wood can and should be used as a design element.

Compare the two panels shown here: They're identical in size, all the frame parts are made from cherry from the same stack, and the panels are both ¼" cherry plywood. The only difference is how the grain was or wasn't used to advantage. The right panel shows care and thought. Notice how the straight grain is similar in the frame pieces; the grain in the plywood tilts in and up, guiding the eye from top to bottom. The left panel shows a total disregard for grain. The haphazard selection of both frame pieces and plywood give the overall piece a disjointed appearance that does not please the eye.

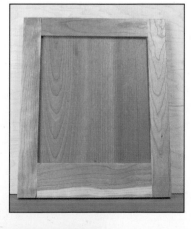

SAVING MONEY

You can save money several ways when buying hardwood. It means spending time to save money; so it all depends which is more precious to you.

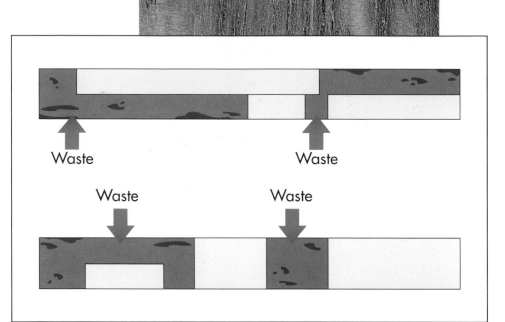

Common boards

Many lumberyards and sawmills don't sell No. 1 or No. 2 common hardwood, for a reason: They don't make any money on it (bad for them, good for you). I've found common lumber that's half the cost of FAS lumber. As discussed in Chapter 4 (*see page* 85), the basic yield of No. 1 common is around 67%; for FAS it's 83%. So for twice the price, you're only getting 16% more usable wood. The tradeoff is time: With common lumber, you'll need to make more cuts to get to the clear stock (*top drawing*). This really isn't an issue, since the first thing you often do with a board is cut it up into pieces. Common stock just takes a little more thought, and a couple of extra cuts.

Shorts

Some lumberyards sell FAS "shorts." These are basically discount cutoffs of longer 12- to 16-foot FAS boards, typically 4 to 6 feet long (*bottom photo*). Technically, this violates the NHLA grading rules, as all FAS lumber must be at least 8 feet long. (These are not to be confused with select "shorts," which have a minimum board size of 6 feet.) If you don't care about length and the price is right, go for it. But if someone tries to sell you these at the same board footage rate as FAS, tell them there is no such thing as FAS shorts and that you want a discount. You can't get it if you don't ask.

Selects

Another way to save money is to purchase select lumber instead of FAS (*top photo*). Basically, a select board is FAS on one side and No. 1 common on the other. These boards cost less and are perfect for furniture parts where only one side will show. And after all, most parts don't show both faces.

Narrow stock

"Narrow" FAS is another common misnomer (*middle photo*). There's no such grade. The minimum width of an FAS board is 6"—period. If you find FAS "narrow" stock at a discounted rate, the wood is good, and the narrowness isn't a problem, then buy it. Here again, if a lumberyard tries to sell it to you at FAS prices, gently inform them of their mistake and try to negotiate a discount.

MICRO-THIN LUMBER

▧ Resawing and planing lumber can eat up valuable shop time. If time is a factor and you're getting ready to start on a project that calls for a lot of thin stock, consider buying "micro" lumber. Micro-thin lumber is sold by many woodworking mail-order catalogs and comes in a variety of woods, both domestic and exotic; thicknesses vary from 1/8" up to 1/2".

In addition to saving time, you may save money with pre-thicknessed lumber, especially in some of the exotics that blunt cutting edges: You won't have to resharpen saw blades and jointer or planer knives. But make no mistake, micro-thin lumber isn't cheap; you pay a premium to have someone else do this work for you.

SELECTING LUMBER

To me, a trip to the local lumber store or sawmill is an adventure—a treasure hunt. You never know what you'll discover in an old pile of wood or a freshly cut log. Some woodworkers regard selecting wood as a chore, which puzzles me. Sure, it's hard work sorting through heavy planks; but the time and energy you spend finding just the right wood will show in the final piece. Even more puzzling are woodworkers who order wood sight-unseen or go to a lumberyard and simply take the top three or four boards off a stack. Selecting boards by chance like this is a sure way to end up with a finished project that shows inattention to detail and a lack of sensitivity to the material (*see the sidebar on page* 123).

Over the years, I've developed a routine for selecting wood that has served me well. I offer it here as a suggested starting point; we all react to wood differently. The best advice I can give you about selecting wood is to develop a trusting relationship with a local lumberyard or sawmill. These businesses take great care in stacking wood so that it remains in top condition. The last thing they want is for a customer to rifle through the stack for one board and then leave it in a mess. To build a trusting relationship, you need to show them that you're sensitive to their needs and that you'll always leave the stack in the same condition you found it—if not better.

First pass

The first thing I do when I begin sorting through a stack is to eliminate warped wood. Any board in the stack with crook (*see page* 96) will likely be evident by comparing it to a straight board next to it. I pick up each board in turn and flip it on its side to sight down the edge and check for cup or bow (*top photo*). If I

suspect the board is twisted, I'll use a pair of winding sticks (*see the sidebar on page* 127). If the board passes these tests, it goes in the "good" pile; everything else is set aside.

Color matching

When I've set aside two to three times as many boards as I need, I go through the pile again, this time looking to match the boards for color. Sorting the boards from dark to light will help (*middle photo*). Hopefully, as you do this, you'll get boards that are similar—they usually end up in a pile in the middle. Color matching may not be that important if you're planning to stain a project, but the closer the boards are in color, the easier it'll be to achieve an even stain.

Check for sapwood

Depending on the wood you're sorting through, sapwood may or may not be a concern. If you're looking for heartwood only, you'll need

to make a decision about boards that are showing sapwood. If it's a lovely board and there's only a little, keep it. If you're losing more than one-fourth of the board to sapwood (*like the top white oak board in the photo at left*), leave it. Some woodworkers prefer wood with both heartwood and sapwood—it's your call.

Best grain

Finally, I inspect each board carefully for grain patterns. This is also the best time to pick out boards for special parts (*see page 122*). In most cases, I'm looking for a wide range of grain. I usually avoid plain-sawn boards; they tend to cup and have wild patterns that are difficult to match. If I've got a lot to choose from, I gravitate to the rift- and quartersawn boards because they move less and are easier to match. To see how they blend together, I prop them next to each other (*middle photo*).

Knots and wane

You may have noticed that checking for knots and wane isn't one of my criteria for selecting wood. There are two reasons for this. First, if you're sorting through FAS lumber and it was graded properly, you won't find much of either. Second, if you're sorting through No. 1 or No. 2 common lumber, try to develop the same mindset that graders use—they don't see the "defects," they see only good wood.

CHECKING FOR TWIST: WINDING STICKS

Whenever I shop for wood, I always take a pair of shop-made "winding sticks" with me to check boards for twist. They're just two pieces of absolutely straight stock about 16" in length. To use them, set one stick on each end of a board as shown. Then squat down so your eyes are level with one stick, and sight over its top edge to the stick on the other end. If the top edges are parallel, there's no twist; if there's deviation, the board is twisted and will surely cause you problems.

GENERAL STORAGE

Don't let your carefully selected wood degrade because of improper storage. First and foremost, lumber should always be stored off the floor (*see pages 129–133 for rack ideas*). If you store lumber on a floor, especially cement, moisture may seep into the wood; at worst it can cause warpage and decay.

Allow to acclimatize

Whether your newly purchased lumber is air-dried or kiln-dried, you must allow it to acclimatize to the conditions of your workshop before you work on it. This is a common mistake that causes unnecessary aggravation. We know that wood constantly moves and adjusts to come to equilibrium with its surroundings. If you don't allow for this, the wood will continue to adjust as you work it—joints that once fit perfectly may tighten or loosen. As a general rule of thumb, I allow wood to rest for a month in the shop before I work on it (*top photo*).

Sticker if necessary

If the initial moisture content of the wood is high, say above 12%, you'll need to allow more time for it to acclimatize. You can help the process along by re-stickering the wood in the shop (*middle photo*). The stickers will allow better air movement and help speed drying.

Store vertically if necessary

If storage space is tight, you can temporarily store lumber by stacking it vertically (*bottom photo*). Make sure the wood doesn't rest directly on the floor, though. A scrap resting on plastic will help keep out moisture. Also, try to stack the lumber as vertically as possible; elastic cords or rope can add support as necessary.

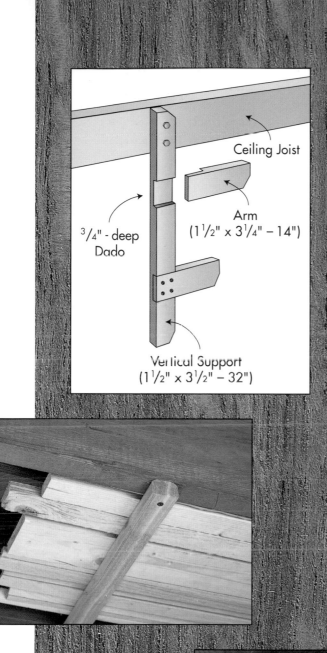

Ceiling Joist

Arm
(1½" x 3¼" – 14")

³/₄" - deep
Dado

Vertical Support
(1½" x 3½" – 32")

OVERHEAD STORAGE

There are several ways to store wood in the unused overhead space in a basement, garage, or shed: a simple storage rack, cleats to hold cut-offs, and short sections of plastic pipe to hold dowels or molding.

Storage rack

The overhead storage rack (*top drawing*) has three parts: a vertical support and two arms. Three or four of these racks attached to floor or ceiling joists with lag screws or bolts will hold a surprisingly large amount of wood. The rack parts are cut from 2-by material and are joined together with T-half-laps strengthened with glue and screws. Since the racks hang from the ceiling, I averted nasty head bumps by knocking off the sharp corners on the bottom of the arms and the top and bottom of the vertical supports.

Cleats

The space between ceiling or floor joists is perfect for storing cutoffs. All you need to do is span the joists with cleats screwed at regular intervals (typically 16"). To prevent the ends of the cutoffs from catching on the cleats as they're slipped in and out, rout a ½"-roundover on the edges of one face of each cleat. Then attach each cleat to a joist so the rounded edges face up, as shown in the middle photo.

Plastic pipe

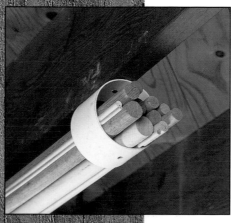

I like to store small or delicate wood parts, like dowels or inlay, overhead where they're out of harm's way. To do this, I cut short lengths of plastic pipe (commonly used for waste or drain lines) and screw them to ceiling or floor joists as shown in the bottom photo. You'll need to drill a pilot hole in the top of each piece for a screw, and a larger access hole in the bottom for the screwdriver.

LUMBER RACK

Although the lumber rack shown below is designed to be built as a stand-alone unit that can attach to any wall, it can also be quickly made using the existing exposed studs in your basement, workshop, or shed. Lumber is stored on short lengths of black iron pipe that fit into holes drilled in the vertical pieces or wall studs. (Note: Before you drill a series of holes in wall studs, check with a local building inspector to make sure the wall isn't load-bearing and that drilling holes won't weaken it.)

Even the space between the verticals or studs is used for storing cutoffs; the cutoffs are held in place with elastic stretch-cords that hook into metal eye hooks installed in the sides of the studs. A length of 1×4 across the bottom of the rack supports the bottom of cutoffs stored, and adds stability if the lumber rack is built as a separate unit. Two more 1×4s are notched into the top and bottom back of the stand-alone rack to provide additional support. All other parts butt together and are joined with screws.

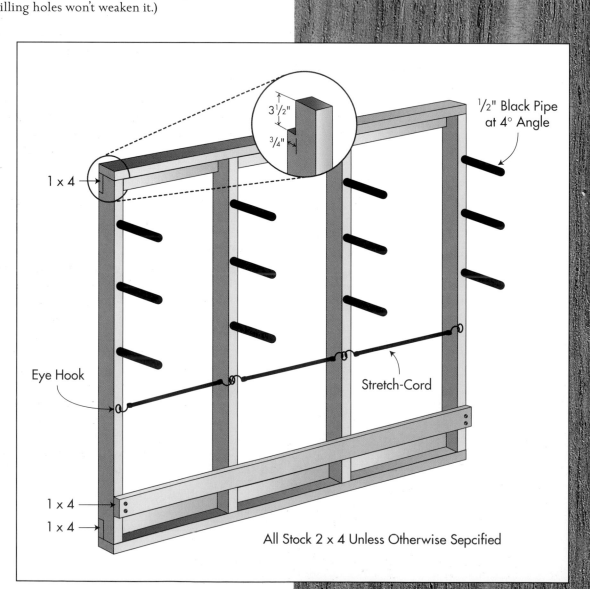

1 x 4

3¹/₂"

³/₄"

¹/₂" Black Pipe at 4° Angle

Eye Hook

Stretch-Cord

1 x 4

1 x 4

All Stock 2 x 4 Unless Otherwise Sepcified

1/2" Black Pipe,
15" Long

4°

7/8"-dia.
Hole,
3" Deep

The trick to building this lumber rack is drilling accurate angled holes for the black pipe (*see drawing at left*). First, to make sure the holes for each "shelf" align. use a long level to mark the hole locations in the studs. To drill the angled holes, I made a simple drilling guide. It's just a 4"-long piece of 2×4 with a 4"-square piece of 1/4" hardboard fixed to the side for clamping the guide to a stud. You can either drill a 4° hole through the width of the 2×4, or drill a straight hole and cut a 4° taper on the back edge of the 2×4. A mark on the side of the hardboard indicating the centerpoint of the hole will make it easy to align the guide to the marks you made on the studs.

Black pipe

Black pipe is sold in various lengths at most building centers (*middle photo*). I found 30"-long sections and had these cut in two (or you can cut them yourself with a hacksaw). After the pipe is cut, round over the ends with a mill file or on a grinder. Black pipe is often grimy, so make sure to clean it thoroughly with mineral spirits before storing wood.

Elastic cords

Elastic stretch-cords provide the most convenient access to cutoffs stored between the studs (*bottom photo*). For 16" on-center studs, a 14" cord is just right—you can purchase these at almost any hardware store or home center.

The elastic cords hook into eyes screwed into the sides of the studs. For maximum flexibility, you can install a couple of different levels so you can move the cords up or down to match the length of the cutoffs.

Sheet Goods Rack

The sheet goods rack shown below is the perfect companion to the lumber rack shown on page 130. It attaches to the wall with hinges, and pivots out on casters for easy access to the stored sheet stock. It can hold up to five ³/₄"-thick 4×8 sheets, or a variety of other thicknesses, as well as sheet good cutoffs.

Construction

You can make this rack with a single sheet of ³/₄" plywood and a couple lengths of 2×6. Start by ripping the plywood in half lengthwise.

Then clamp the two pieces together and lay out and cut the taper on the front ends with a circular saw. Level out the rough cuts with a belt sander or a hand plane.

Next, make the L-shaped frame by screwing a 2-foot length of 2×6 to the end of an 8-foot piece. Now sandwich and clamp the L-shaped frame between the plywood sides. Screw the sides to the frame every 8" or so with wood screws. To help prevent the edges of the plywood sides from splitting and tearing, I cut a ¹/₈" chamfer along all exposed edges with a block plane (or you could use a sanding block). It's also a good idea to round-over the top front edge of the bottom 2×6 to make it easier to slide sheet stock into the rack.

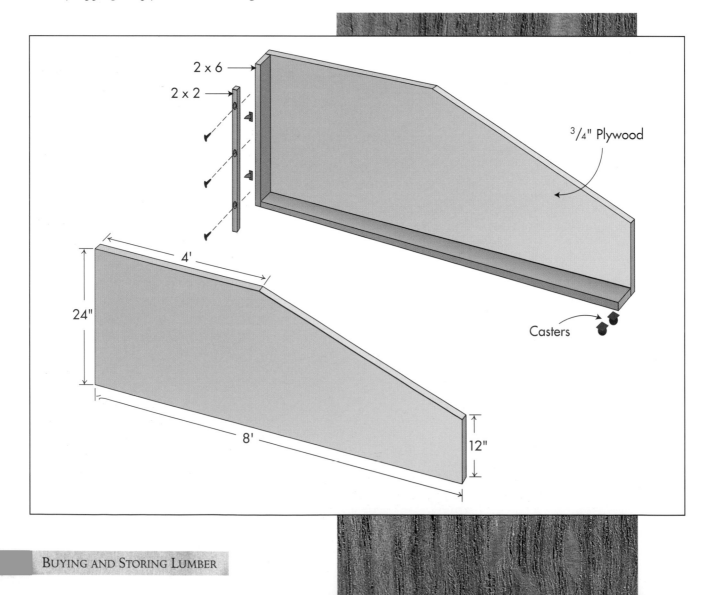

2 x 6

2 x 2

³/₄" Plywood

Casters

4'

24"

8'

12"

If you're adding the sheet goods rack to the lumber rack shown on page 130, you don't need to make and attach the mounting cleat shown at left. Instead, you can attach the hinges directly to a wall stud.

Attach mounting cleat

The easiest way to attach the sheet goods rack to a wall is to use a mounting cleat (*top photo*). This is nothing more than a 2-foot length of 2×2 with counterbored holes drilled in it for lag screws. For a masonry wall, you'll need to drill holes with a masonry bit to accept lag shields. On framed walls, locate a wall stud with a stud finder, and drive lag screws through the cleat directly into the stud. In either case, make sure the cleat is plumb and set above the ground to match the height of the casters.

Add casters

Flip the plywood rack so the bottom faces up. Then install two casters about 2" in from the end of the 2×6 (*middle photo*). Either fixed or swivel casters will work here; just make sure they're heavy-duty and are rated to support considerable weight—5 sheets of MDF weigh in at over 500 pounds. Use beefy screws at least 1" in length.

Install hinges

The best way to support the rack is with strap hinges (I used 5" hinges). You can either attach the hinges to the cleat before attaching it to the wall, as I did, or attach them after the cleat is mounted. Position the rack against the wall and insert a scrap of wood under the back so the rack is level. Then use beefy screws at least 1" in length to secure the hinges to both the cleat and the rack (*bottom photo*).

> "Plywood, with a 3,500-year performance record, is the pinnacle of wood engineering—no other product improves wood's natural characteristics as greatly."
>
> J. F. BURRELL (1972)

PLYWOOD

Although humble and common, plywood has a remarkable history. Since the time of the pharaohs, humans have recognized the benefits of gluing together thin strips of wood to achieve lightweight strength plus stability. In ancient Egypt, plywood formed the cases that entombed mummies. Much more recently, Americans enlisted the properties of this versatile material in World War II: Hardwood plywood was used to build airplanes, boats, and barracks. That's quite a history for something so simple: Plywood is three or more layers of wood glued together so the grain directions are 90° to each other.

Today, plywood is a woodworker's dream: It's flat, stable, and strong. Large panels can be purchased without the trouble of gluing up separate boards. As with all wood, forewarned is forearmed, and I was neither the first time I used plywood. Ignorant of the difference in grades, I bought a sheet of exterior plywood (it was the cheapest) for a sewing cabinet I was planing to veneer. Who knew how poorly its uneven surface would serve as a substrate for veneer? With all the different types of plywood available, an informed choice can make even the humblest project a fine piece of work.

CONSTRUCTION

Traditionally, plywood has been made by gluing together sheets of thin veneer so the grains of successive sheets are perpendicular to each other. This cross-ply construction does two things. First, it creates strong panels—pound for pound, plywood has been proven to be stronger than steel in static bending strength. Second, cross-ply construction creates stable panels.

Consider how much an 8-foot-wide panel made of glued-up boards will change in width as humidity changes. The general rule of thumb is $1/8$" per foot. This means it could move 1" in width throughout the year! But since the grain of the veneer in plywood runs in opposite directions, and the glue that holds the plies together is stronger than the wood itself, there's hardly any movement at all—typically less than 0.02%.

Strength and stability aside, plywood confuses many woodworkers because some of it isn't really plywood. When we think of plywood, we envision an inner core that's made of cross-plies of veneer. Although veneer-core is the most common type of plywood, "plywood"

is also manufactured with an engineered-panel core (particleboard or medium-density fiberboard) or with a solid-lumber core (*see the drawing and chart below*). In some cases, you can even find a combination of core materials. This type of plywood, commonly known as multi-core, mixes layers of engineered wood with veneer.

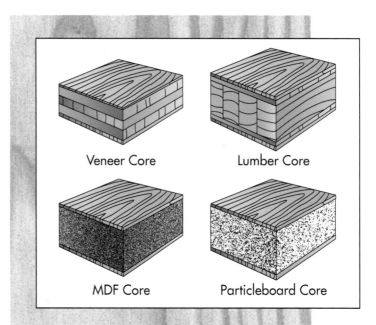

Veneer Core

Lumber Core

MDF Core

Particleboard Core

CHARACTERISTICS OF PLYWOOD

Core Type	Flatness	Visual Edge Quality	Surface Uniformity	Dimensional Stability	Screw-Holding	Bonding Strength	Availability
Particle-Board	Excellent	Good	Excellent	Fair	Fair	Good	Readily
MDF	Excellent	Excellent	Excellent	Fair	Good	Good	Readily
Hardwood Veneer	Fair	Good	Good	Excellent	Excellent	Excellent	Readily
Softwood Veneer	Fair	Good	Fair	Excellent	Excellent	Excellent	Readily
Lumber	Good	Good	Good	Good	Excellent	Excellent	Limited
Hardboard	Excellent	Excellent	Excellent	Fair	Good	Good	Readily

NOTE: Table developed by the Hardwood Plywood & Veneer Association in cooperation with the Architectural Woodwork Institute

Lumber core

The first thing that's important to realize about lumber-core plywood (*top photo*) is that the core can be either softwood or hardwood. However, mixing species within the core is not allowed. There are three grades of wood strips used to make up the cores: clear, sound, and regular. Clear-grade strips must be full-length or finger-jointed and free from knots or other defects. Sound grade is similar to clear, except that it allows discolorations and sound knots. Regular grade allows strips to be butted together without a finger joint.

MDF or particleboard core

Even though it has been around for years, many woodworkers (including myself) still don't think of MDF (medium-density fiberboard) or particleboard plywood as plywood. In reality, it's just an engineered panel with veneer applied to both faces (*middle photo*). Although it doesn't offer the screw-holding ability and strength of standard plywood, it does have other advantages. Because the surface of particleboard and MDF is so flat, the surface of the plywood made with these as cores is very uniform and the panels are extremely flat.

Veneer core

Although MDF and particleboard cores are gaining in popularity, veneer-core plywood is still the standard (*bottom photo*). The main reason for this is the superior strength and screw-holding characteristics that can be found only with this type of plywood. One note about veneer-core plywood: The thickness will vary from manufacturer to manufacturer. Always measure the thickness (preferably with dial calipers) before cutting any joinery. In many cases, 1/4" plywood won't be exactly 1/4"; if you cut a 1/4" groove for it to fit into, you won't get a tight fit.

HARDWOOD FACE VENEERS

How the face veneer for a sheet of plywood is sliced will have a great impact on its overall appearance. The four most common types of slicing are rotary, plain, quarter, and half-round.

Rotary

Rotary-sliced veneer is produced by pressing a broad cutting knife set at a slight angle against a rotating log (*top drawing*). The log is placed between centers of what is basically a huge lathe. The veneer is peeled off the log as if it were a giant roll of paper towels. Rotary-cut veneer can be sufficiently wide to provide a full sheet or a one-piece face.

Plain-sliced

When veneer is plain-sliced or flat-cut, the knife cuts parallel to the center of the log (*middle drawing*). In most cases, the log is brought down against a stationary knife. The innermost growth rings of the log often produce what is known as a "cathedral" grain effect.

Quarter-sliced

Quarter-slicing a log results in straight, uniform grain. One-fourth of a log is brought down against a stationary knife to slice pieces perpendicular to the annual growth rings (*inset drawing*). In some species, like white oak and sycamore, this produces a distinct ray fleck or silver grain pattern.

Half-round

The motion in half-round slicing basically is a combination of rotary and plain slicing. The log is moved in an abbreviated arc roughly parallel to the center of the log (*bottom drawing*). This technique creates the more attractive patterns of plain slicing while getting the most out of the log. The resulting veneer is similar to plain slicing except that the "cathedral" patterns have more rounded tops.

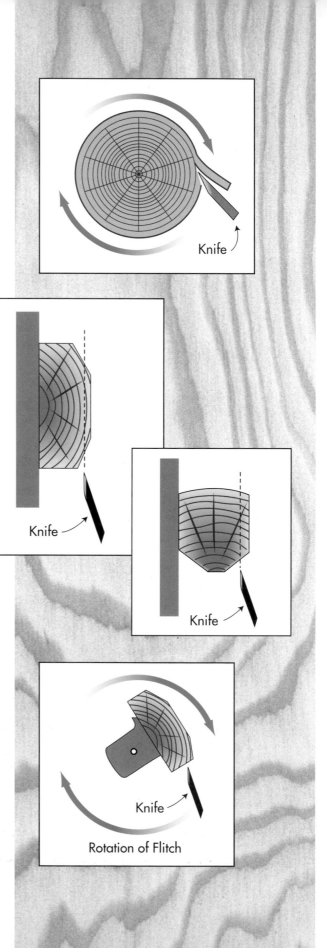

Knife

Knife

Knife

Knife

Rotation of Flitch

Unless veneer has been rotary-sliced to produce full-sized or one-piece sheets, the pieces of sliced veneer must be assembled into full sheets. How the pieces are assembled will also affect the finished appearance of the plywood. There are four main patterns that can be formed by matching sliced veneer in different ways: book-match, slip-match, random match, and pleasing match. Pieces of veneer are spliced together temporarily on a taping machine or tapeless splicer before they're glued to the inner plies.

Book-match

When alternating pieces of veneer sliced from a log are turned over so that adjacent edges meet, the pattern created is referred to as a book-match since the pieces resemble the opened pages of a book (*top drawing*). The growth rings of the wood match up to form interesting symmetrical patterns. This method of matching yields the maximum continuity of grain.

Slip-match

A slip-match pattern is created when successively cut pieces of veneer are slipped out in sequence and joined together (*middle drawing*). This produces a repeating figure where the grain doesn't match up at the joints. Slip-matched plywood is typically very uniform in color, as all faces have similar light reflection.

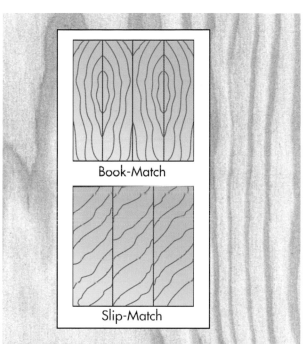

Book-Match

Slip-Match

HOW MATCHING AFFECTS APPEARANCE

Book: When successive pieces of sliced veneer are opened like a book and joined together, a symmetrical pattern of continuous grain is created.

Slip: If the pieces of sliced veneer are slipped off the stack and joined together in order, it creates a repetitive pattern with no continuity of grain.

Random: Pieces of veneer that are joined together at random create a jumbled pattern. Pieces may or may not even come from the same log.

Random match

On plywood faces that are random-matched, no attention has been paid to the sequence or placement of veneers (*top drawing*). This is typically done only on lower grades of veneers, where knots, stains, and other natural characteristics are allowed. Quite often the pieces of veneer are of different widths. This creates a casual "boardlike" effect similar to planks that have been glued up at random.

Pleasing match

When a pleasing match is created, care is taken to match up the color of each of the veneer pieces, but not necessarily the grain (*middle drawing*). This results in a pattern where there is no color contrast at the joints. A pleasing match is basically a combination of book- and slip-matching and random matching. Pleasing-match and random-match faces are most often used as back veneers.

Random Match

Pleasing Match

CHARACTERISTICS OF COMMON WOODS

Species	Plain-Sliced	Quarter-Sliced	Rift-Sliced	Rotary-Sliced
Birch	yes	no	no	yes
Cherry	yes	yes	no	yes
Lauan	no	yes	no	yes
Mahogany	yes	yes	no	yes
Maple	yes	yes	no	yes
Oak, red	yes	yes	yes	yes
Oak, white	yes	yes	yes	yes
Walnut	yes	yes	no	yes

AT THE PLYWOOD MILL

Before being loaded onto a lathe or slicing machine, all veneer logs are steam-heated to soften the wood fibers and assure a smooth texture and easier cutting.

Lathe and charger

Some 80% to 90% of all veneer is cut by the rotary lathe method. As the lathe spindles move (*top photo*), the log is rotated against a knife. The speed with which the knife and the knife carriage move toward the center of the log regulates the thickness of the veneer. Newer lathes use powered rollers to help the lathe spindles turn. An even more recent development is the high-tech spindle-less lathe. Without spindles, this lathe is capable of peeling a log all the way down to a 2" peeler core.

Rotary veneer clipper

After veneers are cut, they go directly to a clipper, which trims the veneer to various widths and removes defects (*middle photo*). Basically, the green veneer ribbon passes over an infrared scanner, which identifies voids, splits, and knotholes and then automatically clips them out. Advanced clippers can be programmed to remove sapwood as well.

Rotary veneer dryer

From the clipper, the veneer goes to the dryer (*bottom photo*). The large chambers of the dryer are equipped with heating elements and fans to circulate air. The veneer is transported on a conveyer belt at a set rate to guarantee that the veneer is dried to a moisture content below 12%. This moisture content is compatible with both the gluing process and the eventual end-use environment of the plywood.

Photos courtesy of the Hardwood Plywood and Veneer Association

Slicer

Most decorative face veneers aren't peeled on a rotary lathe. Instead, they're cut into thin pieces on a slicer (*top photo*). With this method, a section of a log or "flitch" is attached to the log bed, which moves up and down against a stationary knife, cutting a slice of veneer with each stroke. Veneer may be plain-sliced, where the knife is parallel to the growth rings, or quarter-sliced, where the knife is perpendicular to the growth rings.

Sliced-veneer dryer

Just like the dryers for rotary-sliced veneer, the dryers for sliced veneer remove moisture to a level compatible with gluing. Here again, the veneer is conveyed automatically through a heated chamber with forced-air circulation (*middle photo*). Smaller mills also kiln-dry sliced veneer. Since sliced pieces are substantially narrower than rotary-cut veneer, sliced-veneer dryers are typically much smaller in size.

Cross-feed splicer

After drying, less-than-full-sized sheets are dry-clipped and joined together to form full sheets (*bottom photo*). The machines that do this are called splicers, taping machines, or tapeless splicers. First, the edges of the thin strips of veneer are coated with hot-melt adhesive. Then these edges are pressed together, and heat is applied to melt the glue and form a temporary bond. The bond needs to hold just long enough for the full sheet to be bonded to the inner plies. Some machines use a fiberglass thread coated with adhesive to join the pieces.

Photos courtesy of the Hardwood Plywood and Veneer Association

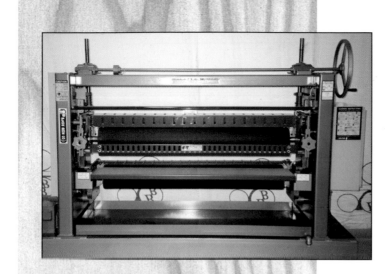

Glue spreader

Once full sheets have been produced, they're ready to be glued up into hardwood plywood. Alternate lower-grade ply veneers for the core are fed through a glue spreader, which simultaneously coats both sides with a thin layer of liquid adhesive (*top photo*). On face veneers, the glue is spread on only one side. The spreader roller controls how much adhesive is applied to the veneer. After the adhesive is applied, the sheets of veneer are "laid up" or stacked into plywood by hand, machine, or a combination of both.

Hot press

Now all it takes to turn the separate sheets of veneer into plywood is heat and pressure. In most plywood plants, the veneer sandwich is first passed through a set of cold rollers to flatten veneers and help transfer adhesive. Then these panels move onto the hot press (*middle photo*). A hot press is a huge machine typically two stories high with dozens of slots in it to accept individual panels. When the press is closed, it exerts anywhere from 150 to 300 pounds per square inch at a temperature of about 250° F. After pressing, the panels are stacked to allow the adhesive to cure and then are trimmed to final size.

Sander

After the panels are trimmed to final size, they're graded. Knotholes and splits on the backs of some grades may be repaired with either wood or synthetic patches. Panels that don't meet specifications are downgraded, remanufactured, or rejected. The final step for most hardwood plywood is sanding. Here the panels pass through wide-belt sanders, where they are sanded to final thickness (*bottom photo*).

HARDWOOD PLYWOOD

Hardwood plywood is available in many different types, grades, and sizes. Many woodworkers think the "quality" of hardwood plywood is better than softwood plywood because it's made out of hardwood. This isn't always the case. The inner plies of hardwood plywood can be made of softwood. It's only the outer or "face" veneers that have to be hardwood. As a matter of fact, often the core of the plywood is a manufactured panel such as particleboard or medium-density fiberboard (MDF). The bottom line is, you're paying a premium for the face veneers.

Types

There are three types of hardwood plywood available: Technical Type, Type I, and Type II. Basically, the difference among the three types is how well the plies stay together when exposed to different conditions. Both Technical Type and Type I are waterproof and are used where the plywood will be exposed to water. Technical Type has the better moisture resistance and is often used in boatbuilding. Since Type II plywood is moisture-resistant but not waterproof, it should be used only for indoor applications—most of the hardwood plywood you'll find in home centers and lumberyards is Type II.

Hardwood plywood grades

The different grades of hardwood plywood describe various characteristics allowed in each veneer or ply. Since the "face" ply is exposed, it determines the appearance of the panel. On some panels, high-grade veneers have been carefully chosen to match color and grain. Other panels show more character marks to give a more natural appearance. The chart below

HARDWOOD PLYWOOD GRADES

Face Grades		Back Grades		Inner-Ply Grades	
AA	Best-quality grade for high-end use	1	Allows sound, tight knots less than 3/8"; no knotholes	J	No knotholes allowed; splits and gaps less than 1/8"
A	Excellent appearance, not as perfect as AA	2	Allows sound, tight knots less than 3/4"; knotholes up to 1/2" if repaired	K	3/8" to 3/4" knotholes allowed, depending on thickness; splits and gaps under 1/4"
B	Used where natural characteristics are desirable	3	Allows sound, tight knots less than 1 1/2"; knotholes up to 1" if repaired	L	1" knotholes allowed; splits and gaps under 1/2"
C/D/E	Used where surfaces will be hidden; C, D, and E allow repairs in increasing sizes	4	Allows sound, tight knots of any size; knotholes up to 4" if repaired	M	2 1/2" knotholes allowed; splits and gaps less than 1"
Special	Appearance depends on species, such as wormy chestnut or bird's-eye maple				

details the different grades in the voluntary American National Standard for Hardwood and Decorative Plywood. There are separate grades for the face, back, and inner plies. It's important to note that the grade assigned to a sheet of hardwood plywood describes only its appearance, not its core type or strength.

Hardwood plywood grade stamps

Since appearance is important, the grade of a sheet of hardwood plywood is stamped on the edge. There are five parts to the grade stamp (*top photo*): the grade (the first character is the face grade, the number is the back grade), species (birch, oak, etc.), construction (number of plies), mill number (may not be present on all stamps), and the standard by which the panel was graded.

Unfortunately, there is no mandatory certification for grade stamps. This basically means a manufacturer can stamp any grade it wants on a panel. Reputable manufacturers, however, will note the Voluntary Standard they followed directly on the purchase order to the lumberyard. If the plywood is of questionable quality, ask to see whether the lumberyard can back up the grade stamp in writing.

Buying tips

I can't tell you how many times I've watched a woodworker sort through a stack of high-grade plywood to find a piece where both faces are near perfect, only to find out the plywood is to be used for a tabletop where only one side will show. The real waste is that A-1 plywood (*top photo*) can cost 20% to 30% more than the same species in A-3 or A-4 (*bottom photo*). Simply put, pay for only what you need. Another way to save money when buying hardwood plywood is to ask whether the store carries any "shop"-grade plywood. This unofficial grade can be either factory seconds or damaged sheets. You'll find these sheets typically priced 25% less than standard sheets.

SPECIALTY PLYWOODS

In addition to standard hardwood plywood, there are a number of specialty and premium plywoods available to handle the woodworking needs of most any project. Premium plywoods such as Baltic Birch and ApplePly, bending plywood, and even prefinished plywood can make a difficult job easy.

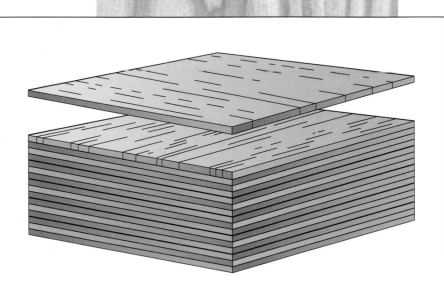

Premium plywood

The big difference between premium plywood and standard plywood is the thickness and number of plies. Basically, premium plywood has more, thinner plies (*top drawing*). This does three things. First, more plies mean greater stability and a much higher strength-to-weight ratio. Second, its screw-holding capability is much better than conventional-core plywood's. Third, unlike the unattractive edges of standard plywood, premium plywood offers a unique, attractive edge that can be cleanly machined.

Jigs

This combination of stability and strength is why I use premium plywood most often for making jigs. Less movement and greater stability means that the jigs are more precise—cuts are straighter, joints are tighter. Baltic Birch plywood is manufactured in Europe and comes only in metric sizes (*middle photo*). It can be hard to find, and when you do locate some, you'll often find only sheets that are roughly 60" square (that's as big as it comes).

ApplePly

ApplePly is the trade name that States Industries has given to their premium, American-made multilayer hardwood plywood (*top photo, page 147*). The core laminations of

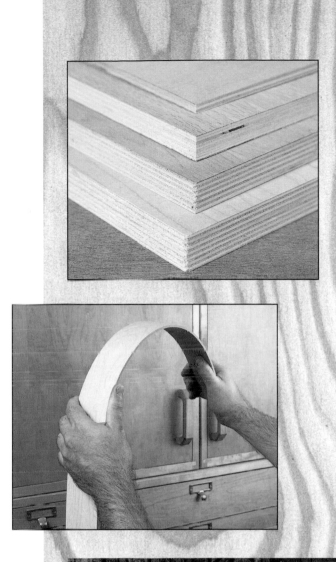

AplePly are $1/16$"-thick veneers of western red alder—this produces a strong but light plywood. Unlike Baltic Birch, ApplePly is available in a variety of species (maple, birch, oak, pecan, hickory, ash, and more) and surface treatments (*see the sidebar below*). Thicknesses of ApplePly range from $1/8$" to 1". Contact States Industries at www.statesind.com to find the nearest ApplePly distributor.

Bending plywood

Unlike most plywood products where you want the panel to be rigid, bending plywood flexes easily (*middle photo*). Bending plywood or bending panels are available in $1/8$", $1/4$", and $3/8$" thicknesses and are manufactured by Danville Plywood Corporation. They're sold under the trade name Curv-A-Board and are great for sheathing a curved corner or other tight radius, such as the end of a kitchen counter or anywhere you'd like to incorporate a gentle curve into a project. The panels can bend because the grain of the thicker face veneers runs in the same direction. The grain of the thinner inner ply runs perpendicular to these to provide some modicum of strength. Curv-A-Board is intended for decorative uses only and should never be used as a structural material.

PREFINISHED PLYWOOD

If you're ever faced with a large project that calls for a lot of exposed plywood, such as built-in bookcases or panels for wainscoting, you might want to consider using prefinished plywood. Prefinished plywood has been used in the high-end cabinetmaking industry for years, but it's popping up more and more in smaller shops where time-saving benefits can easily outweigh the added cost. A wide variety of natural and stained or dyed hardwoods are available, most with UV-cured finishes. Contact States Industries at www.statesind.com for more information.

SOFTWOOD PLYWOODS

Most softwood plywood is manufactured for use in either industrial or construction applications. That's why most standards for softwood plywood deal exclusively with how it must perform in a designated application rather than from what or how the plywood is manufactured. Certain grades of softwood plywood, however, are quite suitable for woodworking projects where appearance isn't critical or the plywood will be used as a base for veneer, laminate, or paint.

Softwood plywood grades

Grade in softwood plywood generally refers to the quality of the veneer used for the face and back veneers (A-B, B-C, etc.); *see the chart below*. Grade can also refer to the intended end use of the panel, such as Sheathing, or Underlayment. The standard that most softwood plywood manufacturers adhere to is Voluntary Product Standard PS 2-92, Performance Standard for

Wood-Based Structural-Use Panels, published by the APA, now called the Engineered Wood Association. *Photos from top to bottom on page 147: A-C, B-C, and C-D plywood. Each is cut in half and one half turned over so you can see both faces.*

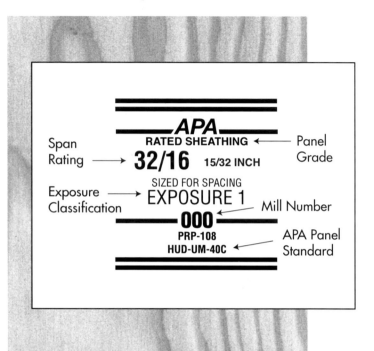

SOFTWOOD PLYWOOD GRADES

Veneer Grade	Characteristics
A	Smooth, paintable. Not more than 18 neatly made repairs, boat, sled, or router type, and parallel to grain, permitted. Wood or synthetic repairs permitted. May be used for natural finish in less-demanding applications.
B	Solid surface. Shims, sled, or router repairs, and tight knots to 1" across grain permitted. Wood or synthetic repairs permitted. Some minor splits permitted.
C (plugged)	Improved C veneer with splits limited to $\frac{1}{8}$" width and knotholes or other open defects limited to $\frac{1}{4}$" × $\frac{1}{2}$". Admits some broken grain. Wood or synthetic repairs permitted.
C	Tight knots to $1\frac{1}{2}$". Knotholes to 1" across grain and some to $1\frac{1}{2}$" if total width of knots and knotholes is within specified limits. Synthetic or wood repairs permitted. Discoloration and sanding defects that do not impair strength permitted. Limited splits allowed. Stitching permitted.
D	Knots and knotholes to $2\frac{1}{2}$" width across grain and $\frac{1}{2}$" larger within specified limits. Limited splits allowed. Stitching permitted. Limited to Interior, Exposure 1, and Exposure 2 panels.

Grade stamps

Reading an APA grade stamp is fairly straightforward. For unsanded panels (*top drawing on page* 148), the panel grade is listed under the APA mark (Sheathing, Underlayment, etc.). Directly under that and to the left is the span rating (32/16), and next to this is the panel thickness ($^{15}/_{32}$"). Beneath that is the exposure durability (Exposure 1), the mill number, and the panel standard. Sanded panels are similar, except the grade is shown as a letter designation (A-C, C-D) and there is no span rating or thickness specified.

Exposure durability

For projects that will be subjected to moisture, such as outdoor play equipment, sheds, or boatbuilding, there are three exposure durability classifications you should be familiar with: Exterior, Exposure 1, and Exposure 2. Exterior panels have a fully waterproofed bond and are designed for applications subject to permanent exposure to moisture. Exposure 1 panels should be used for protected applications where the glue bond must be waterproof. Exposure 2 panels are intended for protected construction and industrial applications. Panels rated as Interior are manufactured with interior glue and should be used only in interior applications.

Quality

Softwood plywood is manufactured from dozens of species of wood, and the quality of the panels can vary tremendously. The face veneer on some plywood is clear and straight; other panels offer wild grain with lots of patches. The quality and number of inner plies used to make the core also varies greatly. It's important that you carefully inspect plywood before buying it. I've seen $^3/_4$"-thick softwood plywood with five inner plies, others with seven.

CHARACTERISTICS OF SOFTWOOD PLYWOOD

There are three common defects to look out for when you shop for softwood plywood: patches, fill, and voids. For the most part, patches and fill affect only appearance; however, voids can create a structural or strength problem.

Patches

Larger knots on A and B grades of softwood plywood can be removed and filled with foot-ball-shaped wood patches (*top photo*). Although this shape is routed to help the patch blend in well with the surrounding wood, it rarely works, because the patch itself has a different grain pattern. Synthetic patching is also becoming commonplace, where the defect is routed out and then a nonshrinking synthetic filler is applied to the void. After the filler sets up completely, it is sanded smooth.

Fill

Smaller knots, gaps, and minor surface imperfections are typically filled with a synthetic wood filler. Voids in the edge of a panel are frequently filled as well (*middle photo*). In some cases, this can be as low-tech as a mill worker using a putty knife to sloppily force filler into the void, or as high-tech as injecting a specialized foam which sets up almost instantly.

Voids

I've always felt voids are the biggest problem to watch out for in softwood plywood (*bottom photo*). Depending on the quality of the inner plies, an edge void may reach in for several inches. This can create a huge problem, especially if you're planning on covering the plies with a wood strip—a void means there's nothing to nail into and not enough surface for a good glue bond. Whenever possible, cut off sections that show voids. If you can't, take the time to fill it with a mixture of 50/50 epoxy and sawdust.

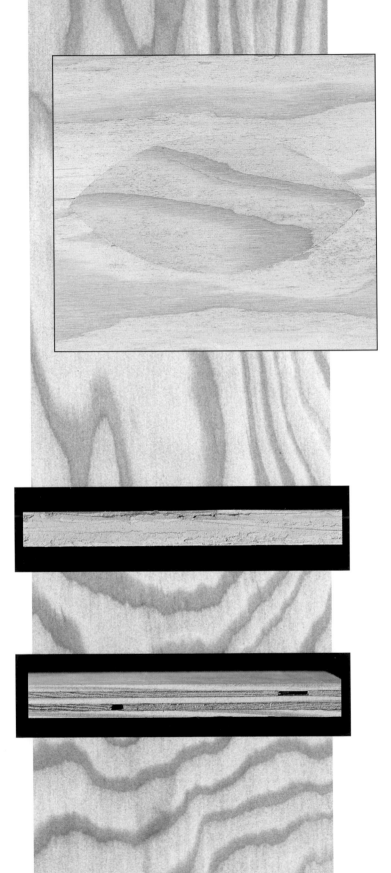

SPECIALTY SOFTWOOD PLYWOODS

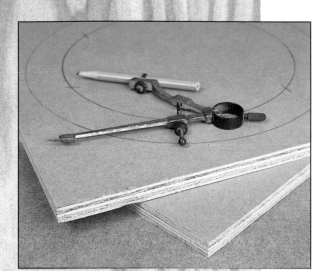

Medium-density overlay

Medium-density overlay or MDO is basically a B-grade plywood that's free of voids and has one or both faces covered with a smooth paper overlay (*top photo*). The overlay is impregnated with a phenolic resin to make the panel moisture- and abrasion-resistant. Although MDO has been used to make road signs for years, I find that it's perfect for making jigs—especially complex ones. The smooth paper face is easy to draw on and resists wear. The lack of voids means the panel holds screws well and will remain stable even with hard use.

Siding

Softwood plywood siding is available with many surface treatments such as V-grooves or channel grooves, and with brushed or rough-sawn textures. Although I don't use siding like the T1-11 shown in the middle photo for many woodworking projects, I've used it for years to cover the interior walls of my shop. Sure, it's more expensive than drywall, but it looks great, and because it's plywood, you can screw into it anywhere. I have, however, used it occasionally for the back of a country-style cabinet to give the piece a rustic feel.

Pressure-treated plywood

Pressure-treated lumber has been treated with a preservative—typically chromated copper arsenate (CCA), which produces the familiar green tinge (*bottom photo*). I don't know of a woodworker who would build a cabinet or a piece of furniture out of it. But it certainly has its uses. I've often used it in air-drying lumber to cover the stack. And it's great for building kilns or other projects where moisture is a concern (like a steam box for bending wood).

"On average, 63% of a tree can be used to make solid lumber. When engineered wood and other products are made from the remaining wood, more than 95% of a tree can be made into useful consumer products."

U.S.D.A. FOREST SERVICE (1990)

ENGINEERED PRODUCTS

Humans couldn't make wood any better-looking, so we settled for added strength, stability, and workability—plus lower cost. Still, many woodworkers dismiss engineered wood products, and that's a shame. This man-made material is not only good for our planet—it helps us conserve natural resources by using more of the tree with less waste—but also a boon to our projects. Using waste like planer shavings that lumbermills used to burn, engineered-product manufacturers produce material boasting stability, flatness, smooth surfaces, and vibration-dampening weight. Your entertainment center, dining room table, and kitchen countertops are probably all made of engineered wood, as is the majority of commercial furniture.

Do these natural/synthetic mixes have a place in woodworking? Absolutely. Are they all pretty much the same? Absolutely not. Just as with solid lumber, knowing what products are available, how they're graded, and how they machine can help you save time and money. No one ever confuses particleboard with pine, or MDF with mahogany; color and grain are still best created by Nature's hand. But when you want workhorse performance on a wage earner's budget, these products are engineered for success.

PARTICLEBOARD

Particleboard is a wood panel product that's produced mechanically: Wood is reduced into small particles, adhesive is applied to the particles, and then heat and pressure turn a mat of particles into a panel product. Particleboard was first developed in the United States and Europe in the 1930s. While a wood shortage in 1940 spurred Germany into early commercial production, wood was still plentiful in America at the time. That's why particleboard wasn't produced here in quantity until the 1960s. In America, it was developed primarily to make use of large volumes of mill by-products—sawdust, planer shavings, etc.

Characteristics

Particleboard is heavy (a $3/4$"-thick sheet weighs almost 100 pounds), flat, fairly stable, and inexpensive. Although there are a number of grades available (*see the chart at right*), you'll probably find just two of these at your local lumberyard or building center: Underlayment and Industrial. Grade stamps like the one shown in the top photo on page 155 provide the grade name, mill number, and certification agency information.

Underlayment is used in flooring; its coarse particle surface makes it unsuitable for most woodworking. The stuff you're after is Industrial grade. Industrial-grade particleboard has a core made of coarse particles sandwiched between two outer layers of finer particles. Finer particles on the outer layers increase the panel's strength and create a smoother, flatter surface that's ideal as a substrate for laminate and veneer (*see page 155*). Particleboard is commonly available in $1/2$", $5/8$", and $3/4$" thicknesses in 4-ft.× 8-ft. sheets. You can also special-order other sizes. Most manufacturers can handle widths from 3 ft. to 9 ft. and any length that's transportable. Thicknesses range from $1/4$" to 2".

Cautions

There are a few important things to know about particleboard. First, although it is fairly stable, particleboard is still made of wood and it can and will react to changes in relative humidity. But it's not linear expansion—less than 1%—that you have to worry about here. The problem is that the wood particles can absorb moisture, which can impede gluing and finishing. For example, I've had laminate peel right off particleboard that was glued up on a humid summer day. Second, unless you're using particleboard made with exterior glue, take care to keep it dry.

GRADES OF PARTICLEBOARD

Grade	Application
M-1, M-S	Commercial construction, such as shelving and kitchen cabinets
M-2, M-3	Industrial construction such as tabletops with high-pressure laminates, case goods, hardwood veneered panels, countertops, and sporting goods
H-1, H-2, H-3	High-density industrial uses where a smooth finish is required
LD-1, LD-2	Core material for solid-core flush and raised-panel doors
M-1, M-2, M-3 –exterior glue	Exterior construction
H-1, H-2, H-3 –exterior glue	High-density exterior construction, such as road signs
PBU	Underlayment under carpeting or other floor covering in residential or commercial 2-layer floor construction
D-2, D-3	Used as a single-layer floor system in manufactured-home construction

Particleboard will wick up liquid water and can swell up to almost twice its thickness. Finally, because of its high resin content, special precautions should be observed when machining particleboard (*see pages 160–161*).

Countertops

If you have a laminate-covered countertop in your kitchen, odds are it has a particleboard core (*second photo from top*). Particleboard is a natural here: It's heavy, flat, and inexpensive. If you're planning to install your own countertop with a bathroom or kitchen sink, try this trick: Before you install the sink, wipe on two or three coats of spar varnish or exterior paint to the edges of the sink cutout. This will seal the particleboard and help prevent it from expanding if it gets wet.

Substrate

As evidenced by its widespread use in laminated countertops, particleboard makes a great substrate for surface treatments (*third photo from top*). Laminate, veneer, and melamine (*see below*) all lay down flat and smooth. Whenever you're applying your own surface, take the time to apply a similar treatment to both sides of the particleboard. If you don't, the unfaced side will absorb more moisture than the other side. If the panel is free to move, it will warp.

Melamine

In addition to veneer, particleboard is available in a variety of surface treatments including melamine, one of my favorites. Melamine is particleboard with a thin layer of plastic applied to both sides (*bottom photo*). A lightly coated plastic surface is useful in many situations, since it needs no finishing. I've used $1/4$"-thick melamine for years for drawer bottoms, particularly in bathroom and kitchen cabinets. The kitchen cabinet industry uses $3/4$" panels to build the actual cabinet cases and then irons on melamine edge-banding. Here again, the beauty is that no finishing is needed. Melamine will stand up well to frequent washing and heavy use.

AT THE PARTICLEBOARD PLANT

The most common process for manufacturing particleboard is the "mat-formed" process shown here. The size of the particles will depend on the end use of the panel and on the manufacturer. For a graduated board, particle size can vary somewhat. To make three-layer boards, the core particles will be longer and the surface particles shorter, thinner, and smaller.

Raw material

Regardless of the desired particle size, the first step in the process is to reduce the logs to chips. Mills use a combination of chippers, hammermills, ring flakers, and ring mills to create these. From here, the chips are further reduced by refiners. Particles are then sorted into like sizes with air streams or screens. Screens are the most common and may be wire cloth or plates with holes or slots.

Drying

The next step, drying, is one of the most critical. To achieve a good bond during forming, the moisture content of the particles needs to be around 2% to 7%. There are three types of dryers commonly in use: rotary, disk, and suspension. In most cases, wet particles enter the high end and are discharged at the low end when dry.

Blending

Dry particles move on to the blender, where they're mixed with binders and other chemicals. The most common resin for particleboard is urea-formaldehyde. The resin content can range anywhere from 6% to 9%. Besides resin, paraffin or microcrystalline wax emulsion is added to help with moisture resistance.

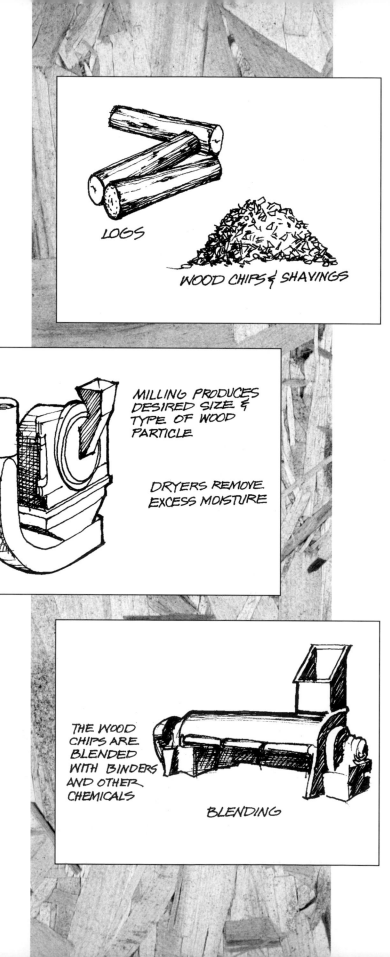

LOGS

WOOD CHIPS & SHAVINGS

MILLING PRODUCES DESIRED SIZE & TYPE OF WOOD PARTICLE

DRYERS REMOVE EXCESS MOISTURE

THE WOOD CHIPS ARE BLENDED WITH BINDERS AND OTHER CHEMICALS

BLENDING

FORMING MACHINES DEPOSIT TREATED PARTICLES ONTO BELTS/ FORMING MATS

PARTICLE MATS ARE COMPRESSED AND BINDERS CURED IN HEATED HYDRAULIC PRESSES WITH TEMP. UP TO 400°F. AND PRESSURES UP TO 1,000 PSI.

CURED BOARDS ARE TRIMMED AND SANDED IN HIGH SPEED BELT SANDERS

SANDING

Forming

Once the particles are blended, they are laid down into an even and consistent mat so they can be pressed into a panel. Most mats are cold-pressed prior to moving onto the hot press, to reduce mat thickness and help consolidate the mat. Three-layer boards require three or more forming stations to lay down the multiple layers.

The hot press

After the mats have been formed and prepressed, they move along to the hot press. There are two common types of hot presses is use: platen and continuous. Temperatures for urea-formaldehyde presses range from 280° to 325°F. The pressure exerted will depend on the press and the material—typically, it will vary between 200 and 500 pounds per square inch. Moisture content of the mat is further reduced in the press by about 3%.

Another way particleboard is formed is by the extrusion process. Here, formation and pressing occur at the same time. Reciprocating pistons force the particles into a long, heated dies—basically pairs of roller platens. This process creates particleboard that has different strength properties than with flat pressing.

Sanding

Pressed boards are trimmed to the desired size. Then the boards are passed along to the sanders, where they are sanded to final thickness. In most plants this is accomplished with high-speed, wide-belt sanders. After sanding, the boards are grade-stamped and stacked for distribution.

Illustrations courtesy of the Composite Panel Association

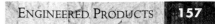

Medium-Density Fiberboard

Of all the composites available, MDF, or medium-density fiberboard, is my favorite (*top photo on page* 159). The reason is how it's made. The process is similar to that used for particleboard—with one important difference: The size of the particles in MDF is much smaller than in particleboard. There's an extra step in the process that breaks wood chips down into tiny fibers. Basically, the wood is cooked in a moderate-pressure steam vessel known as a digester. During this step the wood changes both chemically and physically, becoming less susceptible to the influences of moisture, and less brittle.

In making particleboard, wood is broken apart mechanically; for MDF, though, the wood is rubbed apart into bundles. These fibers are then coated with adhesive (usually urea-formaldehyde), heated, and pressed into a panel. To grasp how much finer the particles are, consider this: It takes roughly a 23"-thick mat of MDF fibers to make a 3/4"-thick panel. Contrast this with the 4" to 6" of wood chips, sawdust, and resin it takes to form the same-thickness particleboard. Breaking down the fibers like this creates the features that I prefer in MDF.

Features

Since there's no grain, changes in relative humidity have little effect on MDF. This means stability—MDF moves less than 0.1% with changes in relative humidity. What's more, finer particles yield a smoother, flatter surface than particleboard. This makes it the perfect substrate (*see page* 159). Uniformly small particles also create a solid, homogenous edge that machines well and holds up over time.

Weight

As you might suspect, packing a lot of fibers in a 3/4"-thick panel results in some serious weight. A standard sheet of 3/4"-thick MDF weighs around 100 pounds. Don't try to carry or cut a sheet of this by yourself—the weight plus awkward size requires a helper. Although a disadvantage when machining, the weight of MDF can be quite useful for damping vibration and creating a stable foundation. It's the perfect base for a tool such as a table saw, drill press, or lathe.

Standard sizes

The standard size for a sheet of MDF is 49" × 97". Since most sheets of laminate are 48" × 96", the extra inch allows a margin of error when you glue the laminate onto the MDF. This way you

GRADES OF MDF

Product Class	Grade	Modulus of Rupture (in MPa)	Modulus of Elasticity (in MPa)	Screw Holding – Face (in N)	Screw Holding – Edge (in N)
Interior MDF	HD – high-density	34.5	3,450	1,555	1,335
	MD – medium-density	24.0	2,400	1,445	1,110
	LD – low-density	14.0	1,400	780	670
Exterior MDF	MD – medium-density	34.5	3,450	1,445	1,110

don't have to worry about aligning the edges perfectly—just lay down the sheet and then use a router with a flush-trim bit to trim the MDF to match the laminate. MDF is manufactured in thicknesses ranging from as little as $5/32$" to as much as $1 5/8$", but you'll most likely find only $3/4$"-thick MDF at your lumberyard or home center.

Homogenous edges

One of MDF's prime features is that its homogenous edges allow for sharp, clean edge-machining with minimal treatment before finishing. In industry, MDF is machined into intricate patterns with little or no fuzzing or chip-out. You can do the same thing in the shop, as long as you use carbide tipped cutting tools (*second photo from top*). (*For more on working with composites, see pages 160–161.*)

Takes paint well

Because the surface of MDF is smooth and dense, it takes paint very well (*third photo from top*). A thin coat of primer and one or two top-coats, and it looks great. (*See page 161 for a tip on dealing with the porous edges of MDF.*) Many of the road signs you see are made from an exterior grade of MDF. The main reason I often use MDF for making children's furniture and storage units is its painting ease—it's easy to get even coats of bright colors with little prep.

Great substrate

MDF is very popular in the furniture industry, where it's the premier substrate for high-grade veneer, thin vinyls, hot-transfer foils, and resin-saturated papers. MDF is dead-flat and supersmooth. Along with premium plywood (*see page 146*), it's the material that I most often use in building jigs (*bottom photo*). I prefer MDF over particleboard as a substrate because the surface is smoother and denser. A denser surface means it absorbs much less adhesive than particleboard: You can usually get a solid bond with a single coat of contact cement, versus the minimum of two required for particleboard.

WORKING WITH COMPOSITES

▓ While composites are mostly wood, their working properties are distinctively different from those of softwood, hardwood, or plywood.

No grain

On the plus side, you don't have to worry about grain direction: There isn't any. This means you'll experience far less tear-out when routing. And there's no such thing as crosscutting or ripping; instead, you just cut the panel.

But this ease of machining has its downside, mostly concerning particle size and the resins used during manufacture. First, the resin is tough on blades. Second, small particles create a superfine sawdust that will quickly line unprotected lungs.

Carbide blades and cutters

You can cut composites with standard steel blades and bits, but they'll dull quickly. Your best bet is to use carbide-tipped bits and blades (*top photo*). Not only do carbide blades and bits stay sharp longer, but they also handle the higher temperatures that often build up when cutting composites. For most cutting, I recommend a 10" combination blade with 50 to 60 teeth and a triple-chip or ATB design.

One way to increase tool life is to reduce feed rates. Since most composites cut so easily, there's a natural tendency to feed the workpiece rapidly into a bit or blade. Although this will cut faster, it'll also quickly dull the blade. Take your time and ease off the pressure. A thin-kerf blade is also a good idea. Removing less material means a cooler blade, which reduces friction and wear.

Dust mask

I once worked in a high-end cabinetmaking shop that used MDF exclusively. Even though I wore a dust mask while machining, by the end of my second day there my sinuses were completely blocked. Yes, the dust collector was on, but the dust from MDF and other composites is so fine, you can't see it. And the particle size is so small, it can penetrate an inexpensive mask. If you're planning on working with composites, do yourself a favor. First buy a quality mask rated for fine particles—such as the cartridge-type shown above. Then keep it on as long as you're in the shop. *(For more on the hazards of dust, see pages 184–185.)*

Biscuit joiner

Although, composites don't have the edge strength required for advanced joinery like box joints and dovetails, they do work well for biscuit joinery. A biscuit joiner, like the one shown in the top photo, cuts a half-oval-shaped groove in both pieces to be joined. The groove accepts a football-shaped "biscuit" made of compressed wood. When glue is injected into the groove and the biscuit inserted, the wood fibers of the biscuit wick up the glue and swell. This effectively locks the two pieces together. As long as you keep the groove as far from the edge as possible, the joint will hold up well over time.

Edge-banding

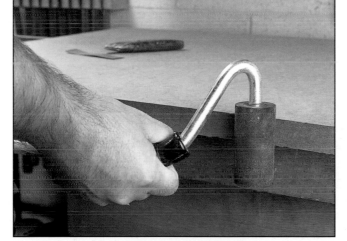

Not all composites have homogenous edges; that is, they may have a denser surface than their core. This creates a problem when finishes are applied, especially a high-end finish where a smooth, uniform high gloss is desired. Basically, the coarser inner core absorbs finish more rapidly—it acts almost like end grain. A simple solution is to apply fiberboard edge-banding to seal the edge (*middle photo*). This can be either self-adhesive or iron-on. I recommend the iron-on banding because you'll get a better, more durable bond.

Special fasteners

Most composites hold screws well and so they are the fastener of choice in the ready-made furniture industry. Modifications have been made to the screws to make them hold even better (*bottom photo*). New and improved composite screws can be purchased from most mail-order woodworking catalogs. For the most part, these screws are threaded the entire length and have deep, open threads to bite into the composite better.

ORIENTED-STRAND BOARD

Oriented-strand board (OSB) is an engineered panel made from strands of wood bonded together with a waterproof resin under heat and pressure (*see pages 164–165 on the manufacturing process*). OSB started appearing commercially in the United States around 1980. Since then it has found wide acceptance in the construction industry, where it's used primarily for roof, wall, and floor sheathing.

Besides using larger pieces than particleboard or MDF, the big difference with OSB is that the strands are laid down so that they're oriented in a certain direction. This creates panels that have cross-banding properties somewhat similar to plywood (*photo below right*).

Exposure classifications

Under the APA guidelines of the Engineered Wood Association (formerly the American Plywood Association), OSB is made with two different exposure ratings: Exterior and Exposure I. Exterior-rated panels have a fully waterproof bond and are designed for use with permanent exposure to moisture. Exposure I–rated panels have a fully waterproof bond and are designed for applications where the panels may be unprotected. Some 95% of OSB is manufactured with Exposure I ratings. OSB is also made in three grades: APA Rated Sheathing, APA Rated Sturd-I-Floor, and APA Rated Siding (*see the chart below*).

Grades and performance standards

Unlike some plywoods and other engineered panels, OSB is rated by a performance standard, rather than a prescriptive standard that defines how the panel must be made. The beauty of a performance standard is, it sets requirements for a product based on the product's intended use. In other words, as long as the panel can do what it's supposed to do, the manufacturer can make it using any materials and methods at their disposal.

Span ratings

Since OSB is a performance-rated panel, you'll always find a span rating as part of the grade stamp. A span rating is the recommended center-to-center spacing of supports, in inches, over which the panels should be installed. For APA Rated Sheathing and Sturd-I-Floor, the span rating applies when the long panel dimension is across supports. The span rating for APA Rated

GRADES OF OSB

Grade	Thicknesses	Application
APA Rated Sheathing	5/16", 3/8", 7/16", 15/32", 1/2", 19/32", 5/8", 23/32", and 3/4"	Can be used for subflooring, wall sheathing, and industrial applications such as shelving, furniture, trailer liners, recreational vehicle floors, roofs, and components
APA Rated Flooring (Sturd-I-Floor)	19/32", 5/8", 23/32", 3/4", 7/8", 1", and 1 1/8"	Intended for use as single-layer flooring under carpet and pad; most flooring has tongue-and-groove edges

Siding panels is for vertical installation.

A span rating looks like a fraction (such as 32/16), but it isn't. The left-side number describes the maximum spacing of supports in inches when the panel is used for roof sheathing; the right-side number denotes the maximum spacing of supports when the panel is used for subflooring.

Although I don't use OSB for woodworking, I do use it for various applications in the shop. Since it's about half the price of plywood yet has some similar strength properties, I use it as a low-cost alternative. The waterproof resin makes this stuff well suited for small outdoor buildings like storage or garden sheds. The two most common uses I have for OSB in the shop are building crates and building mock-ups.

Crates

I've shipped a lot of handmade projects to friends and family over the years in crates made entirely of, or sided with, OSB (*top photo*). If you're planning to make a crate entirely out of OSB, I recommend using at least $5/8$"-thick panels and joining them together with narrow crown staples for better holding power.

Mock-ups

OSB is also a great material to make mock-ups from when designing a new project. Sometimes you just can't visualize the impact of a part or dimension on paper. In these situations, I like to make a full-sized piece out of OSB. If you use at least $3/8$"-thick panels, you can even screw the parts together to get a three-dimensional view. For example, I couldn't quite decide on a length for the aprons of a Queen Anne–style coffee table I was designing. So I built a quick mock-up with two different apron lengths (*middle photo*). It took about 10 minutes and a couple bucks' worth of OSB. I brought it into the house and it was obvious that the shorter aprons would be better. What a work saver!

WHAT AN OSB GRADE STAMP WILL TELL YOU

A certification stamp on OSB provides the following information:

- span rating
- nominal thickness
- exposure durability classification
- grade
- manufacturer's name or mill number
- certification organization logo
- symbol signifying conformance to a performance standard
- quality assurance report number
- direction of surface strand alignment

HOW OSB IS MADE

■ At the start of the OSB manufacturing process, logs are sorted and hauled to the jackladder, where they're presoaked to soften the bark. Then they move up a conveyer to the debarker, which removes the bark and cuts the logs into shorter lengths. The bark is burned as fuel for the mill's power system.

Next, the short logs move into the strander. The strander is a machine that slices the logs into strands along the direction of the grain. Strand dimensions are set by the manufacturer and are cut to a uniform thickness. Most mills use a combination of strands ranging from $3^1/2$" to 6" long and approximately 1" wide.

The strands are temporarily stored in wet bins before moving onto the dryers, where they're dried to a uniform moisture content. Then the strands are sorted and mixed with a waterproof exterior-type binder.

At the forming line, the strands are oriented in layers. The strands on a panel are generally aligned in the long direction of the panel for better bending strength and stiffness in each direction. The two or three inner layers are cross-aligned (like plywood) to the surface layer.

After forming, the mat of strands is pressed at high temperature and pressure to form a rigid panel. The panels are then cooled, cut to size, grade-stamped, and stacked for shipping.

Log Hauling and Sorting

Drying

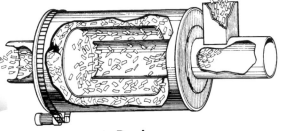

Blending

Forming Line

Illustration courtesy of the Structural Board Association

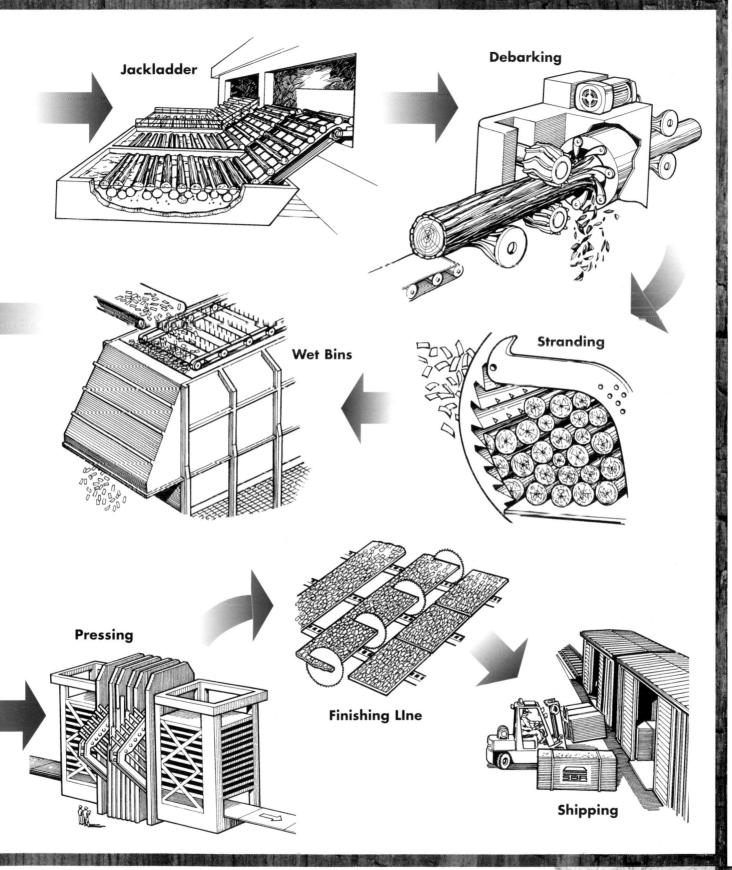

Jackladder

Debarking

Stranding

Wet Bins

Pressing

Finishing LIne

Shipping

HARDBOARD

Another engineered product that's been around for a long time and seen wide use is hardboard. Hard, dense, and relatively flat, hardboard is commonly called Masonite, the brand name of the leading manufacturer. If your house was built within the last 15 years and has siding, odds are it's hardboard. But the most recognizable form of hardboard is pegboard. Who doesn't have a sheet or two of this lining the walls of his or her shop, basement, or garage to organize tools or gardening equipment?

Hardboard is made in much the same way as MDF and particleboard. A mixture of finely ground processed wood and resins are bonded together under heat and pressure. There are two basic types of hardboard, available in two thicknesses ($1/4$" and $1/8$"): standard and service-tempered. Whenever possible, I use service-tempered because it's harder and more resistant to moisture than standard hardboard and has less tendency to delaminate.

Stripes

Although many woodworkers think you can tell the difference between standard and service-tempered hardboard by its color, you can't. The color of the hardboard has to do with the species of wood that was used to make it. Lighter-wood trees produce lighter hardboard. The only real way to tell the difference is to look for stripes on the edges of the hardboard. Two red stripes means it's service-tempered, one green stripe is standard. Since most hardboard is thin, detecting the stripes can be difficult—your best bet is to check the stack of hardboard at the lumberyard.

Waffle or smooth?

The other thing to look for in hardboard is the type that is smooth on both sides. Some hardboard is available with a "waffle" back

(*top photo*). A textured side is a by-product of the manufacturing process when a freshly pressed sheet is transported on a screen that leaves an imprint. I've had consistent troubles getting a good glue bond with textured-back hardboard; stick with the smooth stuff.

Jigs

Since hardboard sheets are thin, I generally use them for drawer bottoms and for jig making. The $1/8$"-thick size is particularly well suited for runners or slides for a jig (*middle photo*). That's because the kerf on a standard carbide-tipped blade cuts a perfect groove to accept hardboard.

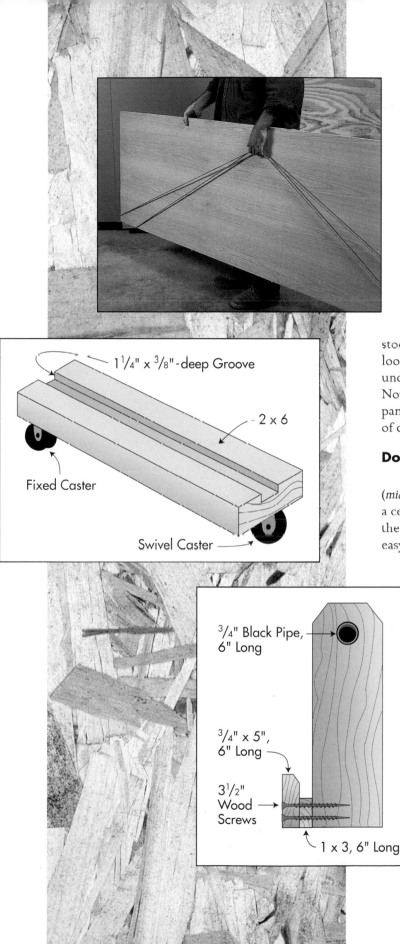

Transporting Sheet Stock

A 4-ft. × 8-ft. panel is a real challenge to move around by yourself. You have to struggle not only with the awkward size, but also with the weight. If the panels are $1/2$" thick or less, try the rope trick or the tote below. For $3/4$"-thick panels, I recommend making a push-around dolly.

Rope trick

You don't have to be a cowboy to use this simple rope trick to transport sheet stock (*top photo*). Just tie a length of rope into a loop. Then place opposite ends of the loop under the two bottom corners of the panel. Now pull up on the rope in the center of the panel to carry it. It'll take some trial and error, of course, to find the right length for the rope.

Dolly

The dolly is a skateboard for sheet stock (*middle drawing*). It's just a 2-ft. length of 2×6 with a centered groove cut in it. Two fixed casters in the back and a swivel caster in the front make it easy to push even the heaviest sheets around the shop. To use the dolly, place a sheet in the dolly's groove so the sheet is centered along its length. Then gently push the panel while keeping it balanced upright.

Tote

Although the rope trick above is good in a pinch, it can be cumbersome to use. A more elegant solution is to extend your arm by building a sturdy tote (*bottom drawing*). The tote consists of two sides held together by an L-shaped cleat plus a length of black pipe. The cleat holds the sheet stock, and the pipe serves as a comfortable, stout handle.

Dolly drawing labels: $1^1/4$" × $3/8$"-deep Groove · 2 x 6 · Fixed Caster · Swivel Caster

Tote drawing labels: $3/4$" Black Pipe, 6" Long · $3/4$" × 5", 6" Long · $3^1/2$" Wood Screws · 1 x 3, 6" Long

ALTERNATIVE MATERIALS

In Mr. Sheraton's time veneers were troublesome, but things have improved considerably. Today's paper-backed, flexible veneers are a joy to work with, and good thing: Some woods are so expensive and rare that the only way most mortals get to work with them is in veneer form.

Due to the increasing cost and scarcity of solid wood in general, it's smart to explore veneer and other alternatives, like recycled wood and even nonwood products. Sometimes, money isn't the issue at all—woods like old-growth pine and chestnut simply don't exist anymore, and they're available only as reclaimed or recycled woods. Don't think, though, that you're "settling" when you choose an alternative. Thanks to advancing technology, some of these materials perform even better than wood.

For example, recycled-wood-and-plastic composites will stand up to the worst elements without sacrificing their looks or strength. Still other alternatives—phenolic, plastic, and laminate—are impervious to changes in humidity, unlike solid wood, so you can use them for more precise jigs and fixtures. When your automatic first choice is natural, solid wood, think again: There are alternatives, and they might be the best materials for the job.

VENEER

Veneer, basically a thin piece of wood sliced from a log, has long been used as an alternative to solid wood. Rare and highly figured pieces were used in post-Renaissance Europe for the finest furniture. But because this type of veneering required expert skills and a lot of patience, it was expensive. When production machinery appeared in the 19th century, the art of veneering declined. But two important things changed that. The first was the invention of the knife-slicing machinery that can produce thin, uniform veneer inexpensively. Second, the development of moisture-resistant and water-proof adhesives swept away the poor reputation that veneered furniture had developed in the 20th century because of frequent delamination.

In the past, veneer was split from a log and painstakingly scraped to a uniform thickness. Today, most all veneer is sliced—rotary-cut, half-round, flat-sliced, or quarter-sliced (*see the chart below*). (*For more on slicing process and techniques, see pages* 138–140.) Typical thickness for veneer ranges from $1/100$" to $1/28$". The thinner veneers are used to make flexible or reinforced veneers (*see the sidebar on page* 171). The thicker

$1/28$" veneer is often referred to as common veneer or natural veneer. Even thicker veneers ($1/10$" to $3/16$") are cut for the manufacture of plywood inner cores.

Knife checks

Although the veneer that's cut with knife-based machines is smooth and uniform in thickness, it does have one problem: knife checks. As the veneer is forced by the knife away from the log at a sharp angle, tiny checks or cracks develop on the knife side of the veneer. This checked

TYPES OF VENEER

Veneer Cut	Typical Species	Standard Sizes
Burl	Carpathian elm, English oak, myrtle, olive, redwood, walnut	irregular dimensions; pieces vary from 6" x 10" to 16" x 52"; typical size is 16" x 24"
Butt and stump	Maple, walnut	irregular dimensions; pieces vary from 12" x 36" to 18" x 52"; typical size is 12" x 36"
Crotch	Mahogany, walnut	width from 10" to 24"; length from 18" to 52"; average size is 12" x 36"
Flat-sliced	Ash, rosewood, cherry, maple, oak, teak	width from 4" to 24"; length 3 ft. to 16 ft.
Quarter-cut	Mahogany, oak, satinwood, zebrawood	width from 3" to 14"; length from 3 ft. to 16 ft.
Rotary-cut	Birch, bird's-eye maple, bubinga, cherry, red oak	width from 8" to 36"; length up to 10 ft.

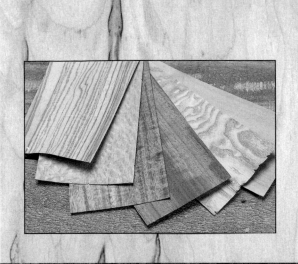

surface can create imperfections on a finished surface long after the veneer has been applied. The checked side is commonly referred to as the loose side; and the opposite side, free from checking, is called the tight side. Whenever possible, the loose side should be bonded to the substrate to prevent future problems.

Sizes

Common veneer is available in a wide variety of species and sizes and is sold by the square foot (*top photo*). Be aware that although most mail-order veneer suppliers will try to meet your request for specific sizes, veneers are cut in random widths and lengths, like hardwood lumber. This means you may not get the exact size you need. Your best bet is to call the supplier to see what they have in stock. Usually, you'll find them more than willing to help.

High-figure veneers

Highly figured veneers like burls, crotches, and butts usually come in small sizes or in preassembled sheets where smaller pieces have been joined with veneer tape. Most of these veneers will arrive far from flat—they're often quite convoluted. Before you can glue them to a substrate, you'll need to flatten them. Most veneering companies sell a special "flattening" liquid that is basically watered-down white glue with a bit of glycose added to lengthen the time before the glue sets. To use it, wipe on a thin coat with a damp sponge and let the veneer rest for a few minutes. Test the edge for flexibility. If it's pliable, lay a plastic garbage bag over it and then carefully set of piece of plywood on top. Then gradually add weights to the plywood until the veneer is flat. Allow it to sit for at least 24 hours.

FLEXIBLE VENEER: PAPER AND PEEL-AND-STICK

There are two types of flexible veneer that have become increasingly popular, and for good reason. Standard and pressure-sensitive adhesive backed (also called peel-and-stick) veneer are easy to work with. And since they are both thinner than common veneer and reinforced with a backing, they can bend around a smaller radius without breaking. This stuff is so flexible that it comes in rolls, typically 24" × 36".

Flexible veneer is so thin that it cuts easily with a utility knife. A word of caution, though: Because it's so thin, it's really easy to sand through the veneer. I don't recommend power sanders for flexible veneer. In the first place, the veneer should not need heavy sanding; secondly, a power sander can damage veneer in the blink of an eye.

Flexible veneer: A paper backing provides support for the thin face veneer and creates a smooth surface that's perfect for use with contact cement.

Adhesive-backed veneer: Commonly referred to as peel-and-stick, this type of veneer has a thin layer of adhesive applied to the back.

MAKE YOUR OWN VENEER

Cutting your own veneer has a number of advantages. It saves money, of course. And by using the same wood as the rest of your project, the veneer you cut will match perfectly. What's more, you can custom cut the veneer to any thickness you want. All it takes is a bandsaw with a sharp blade, a couple of shop-made jigs, and some patience.

Book-matching

One of the things I like best about cutting my own veneer is that it lets me book-match—where consecutive sheets are opened like the pages of a book (*top photo*). When you order veneer from a company, you may not get consecutive sheets.

Bandsaw fence

To cut veneer on a bandsaw, you'll need a tall fence to support the wood during the cut. I recommend making a sturdy fence like the one shown in the middle drawing. It's two pieces of MDF or plywood screwed together at a right angle, with a pair of triangle-shaped support blocks to provide rigidity and ensure a 90° cut. Cutting a notch near the bandsaw's guide assembly lets you lower it when cutting shorter workpieces.

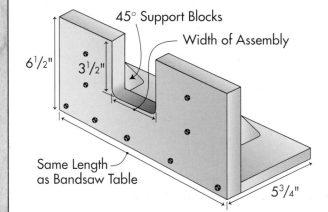

Featherboard

Another simple jig that helps ensure accuracy is a featherboard, which presses the workpiece firmly against the fence (*bottom photo*). It consists of three parts: a base, an arm, and a pressure plate. The pressure plate is a piece of $1/8$" hardboard. In use, the featherboard is clamped to the bandsaw table so the workpiece is pressed against the fence but can move smoothly.

Adjusting the bandsaw

There are a couple of bandsaw requirements that are critical for successfully cutting veneer. The blade must be the right type, and it must be sharp. I use a $1/2$"-wide, 4-tooth-per-inch blade because it's designed for resawing. The gullets on this style blade are deeper, which allows for better waste removal and cooler cuts.

The saw must also be adjusted properly (*top photo*). Make sure the guide blocks and thrust bearing are in the correct position (a gap equal to the thickness of a dollar bill works best). Increase the blade tension to at least the next blade width. For instance, if you're using a $3/8$"-wide blade, adjust tension for a $1/2$". Then lower the blade guard as close to the workpiece as possible, and clamp the fence and featherboard to the table for the desired cut (I generally allow an extra $1/16$" for planing).

Cutting veneer

Turn on the saw and guide the workpiece into the blade *(middle photo)*. Use steady, even pressure. Don't force the cut—let the blade do the work. Remember, you're removing a lot of material here—it's slow going. As you complete the cut, use a push block to safely push the workpiece past the saw blade.

TROUBLESHOOTING: BOW AND ANGLED CUT

The two most common problems encountered while resawing are bowed and angled cuts.

Bow cut

A bowed cut, or "barreling," is usually the result of a blade that's not tensioned or supported properly. First, check to make sure the upper guide assembly is as close as possible to the workpiece. If it is, then increase blade tension until it cuts straight.

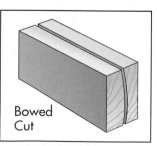

Bowed Cut

Angled cut

In most cases, an angled cut is caused by a table that's not square to the blade. This is an easy fix: Just readjust the table. You might also want to try repositioning the featherboard so it presses the workpiece more firmly against the fence.

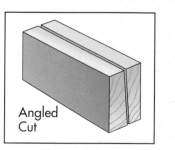

Angled Cut

VENEER ON THE TABLE SAW

Another way to cut veneer is with a table saw (*top photo*). The only limitation is that the maximum width of veneer you can cut will be roughly double the maximum cut your saw can make. What's more, the wide kerf of the saw blade wastes a lot of wood; minimize this waste by using a thin-kerf blade. To safely cut veneer on the table saw, you'll need to make a zero-clearance insert (*see the sidebar below*). This special insert hugs the blade and prevents veneer from dropping down into the normal wider blade slot. A built-in splitter also helps prevent binding.

Two cuts

To make sure the veneer cuts evenly, the saw blade must be perfectly parallel to the rip fence. Raise the blade to cut slightly more than halfway through the workpiece, and position the rip fence for the desired cut. Then, using a push block, make the first cut (*middle drawing*). Here again, use firm, even pressure and don't force it. For dense woods, consider a series of shallower cuts. Now flip the workpiece end-for-end and, with a push block, finish the cut with the second pass.

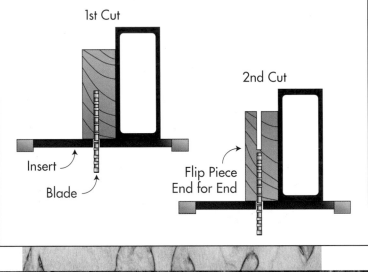

1st Cut

2nd Cut

Insert

Blade

Flip Piece
End for End

ZERO-CLEARANCE INSERT

You can use your standard insert as a template to make a zero-clearance insert. First, place the standard insert on a ¹/₂"-thick piece of plywood and trace around it. Then cut it to rough shape to within ¹/₈" of the outline. Next, temporarily attach the standard insert to the plywood with double-sided tape. Now you can trim it to exact size with a flush-trim bit in a router. Drill a 1" lift-out hole in the insert and cut a kerf in it for a splitter that aligns with the blade. Glue a ¹/₈" piece of hardboard in the end of the kerf, and you're ready to resaw.

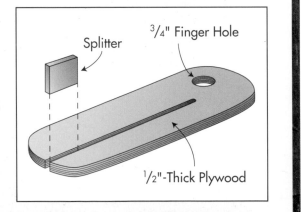

Splitter

³/₄" Finger Hole

¹/₂"-Thick Plywood

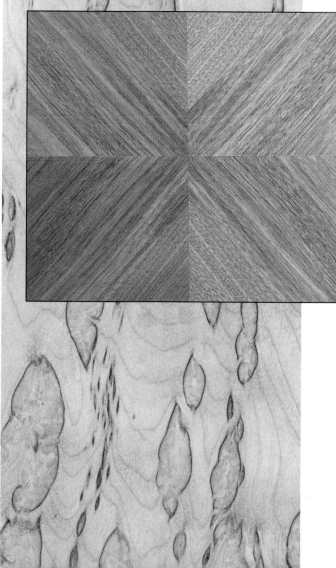

VENEER TAPING PATTERNS

In addition to the standard veneer-matching possibilities shown on pages 139–140, you can achieve several highly decorative patterns with some cutting and taping and a little care and patience.

Quartermatching

Quartermatching produces many fancy patterns. This type of match is commonly used with butt, stump, and crotch veneers, since it's the most effective way of showing them off (*top photo*). The most common quartermatch config-uration is the four-way center and butt match.

Circles, ovals, and other odd shapes are easily obtainable with quartermatching. Small veneered tabletops often use this pattern because it's an excellent way for you to use up short stock that would otherwise be waste.

Diamond

Diamond-matching is a very popular variation of quartermatching. It is espe-cially striking when used with straight-grained wood without much figure. The sheets are cut at an angle and quarter-matched to produce the diamond pat-tern. Reverse diamond-matching is the same principle applied to the same kinds of veneers, but with the grain matched to produce an "X" pattern (*middle photo*).

Matching tips

Without actually cutting the veneer, it's difficult to visualize the patterns you can get with the different matching methods. One trick to get around this is to use a pair of mirrors, approximately 8" wide and 16" long. Butt the mirrors together end to end, and apply a strip of duct tape at the joint to create a simple hinge. Then set the mirrors on a piece of veneer and adjust the angle to get the desired effect.

WORKING WITH VENEER

Whenever you cut your own veneer, you'll be faced with some rough surfaces. Since the veneer is so thin, you can't run it safely through a thickness planer. This is a situation where your trusty old hand tools can get the job done.

Securing the veneer

The first thing you'll need to do is secure the veneer so you can work on it. It's too thin to clamp in place, but double-sided carpet tape will work. Apply a couple of strips to each piece of veneer, and then attach them to a flat base such as a piece of MDF or particleboard (*top photo*). After you're done surfacing the veneer, don't try to pry it off the base—it'll break. Instead, drizzle a little lacquer thinner between the veneer and the base; this will dissolve the tape's glue, and the veneer will lift off easily.

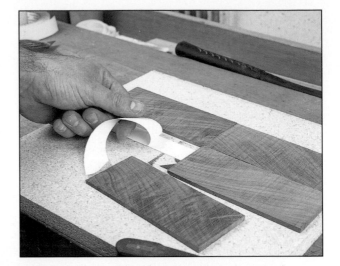

Planing

For woods that aren't highly figured, a well-sharpened jack plane will quickly level the surface (*middle photo*). Adjust the plane for a very light cut, and angle the plane to make a shearing cut. Change direction as needed to follow the grain. Be especially careful of veneer that's quartermatched, as the grain will shift in each of the four sections.

Scraping

If the veneer you've cut is highly figured, I recommend using a scraper to flatten and smooth the surface (*bottom photo*). Although it won't remove material as fast, the tiny burr of a scraper will pose less risk of tear-out. Since scrapers have a tendency to follow saw marks—they ride up and down with them instead of leveling them—it's best to hold the scraper at an angle to the ridges. This way it'll shear off the high point, leaving a flat surface behind.

No matter what the veneer type or how it was cut, there's basically just one way to attach it to a substrate. There are big differences, though, in the type of adhesive you use. If you're applying flexible veneers with paper backing, use contact cement (Unibond 800 brand urea-formaldehyde works great). Do not use contact cement on natural veneers, as it will eventually form loose spots and bubbles in the veneer. White glue works fine, but stay away from yellow glue: The open time is so short, it's impossible to spread it and clamp it before it sets up.

Contact cement

Although you can get away with brushing on contact cement on small projects (*top photo*), it's best to roll it on larger surfaces. I use a 3" or 4" short-nap paint roller. For porous substrates like particleboard, apply at least two coats.

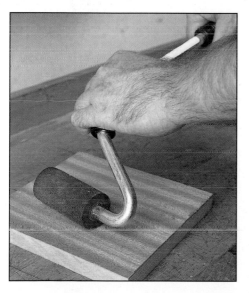

Pressing with a roller

Once the contact cement has set up (it'll gloss over), it's ready for bonding. It's easy to position small pieces without accidentally touching the parts together. Larger pieces require extra precaution. I place dowels every 6" or so on the substrate before placing the veneer on top. Then, working from one end and removing one dowel at a time, I press the veneer firmly onto the substrate with a laminate roller (*middle photo*).

Trimming

As soon as the veneer is attached, any excess can be trimmed off with a utility knife (*bottom photo*) or a veneer saw (*inset*). The tiny teeth of a veneer saw work best, as they don't have a tendency to follow the grain of the wood the way a utility knife does.

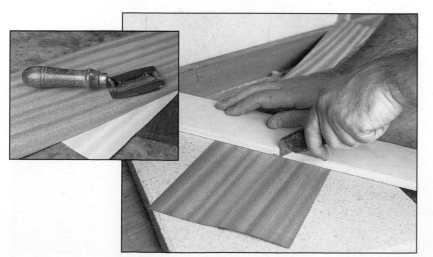

LESSER-KNOWN HARDWOODS

Alternative hardwoods

When most woodworkers think about selecting a wood for a project, usually the old standbys come to mind: oak, cherry, maple, and walnut. It's logical, since this is what's commonly available at the local building center or lumberyard. But with a little extra effort, you can discover a whole new world of alternative woods, exotic or domestic. Many of these not only are as attractive as "standard" woods, but also can be much cheaper. Add to that the fact that many of the alternative exotics are grown in sustainably managed forests (*see the sidebar below*), and you've got a winning combination.

Exotic alternatives

Perhaps the biggest stumbling blocks to using alternative hardwoods are lack of information and, to a degree, the lack of prestige an unknown wood offers. If you tell someone who doesn't know much about wood that the table you just built is made from chechem, how will they react? Probably by asking, "What the heck is that?" But tell that to someone who's wood-savvy, and you'll get a different reaction. "So that's chechem; I've heard about it. How did it work? Any problems?" And so on.

If prestige isn't an issue and you want to break away from the usual, I heartily recommend dipping into the world of alternatives. Shown from left to right are some of my favorites: chechem, chakte kok, jarrah, jatoba, and imbuya. (*For a detailed description of each of these woods, see Chapter 2.*) Each of these looks great and is interesting to work with.

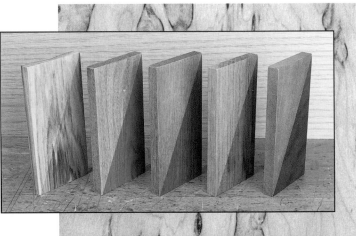

CERTIFIED FOREST PRODUCTS COUNCIL

If you're interested in locating certified wood and wood products, the Certified Forest Products Council (CFPC) website, www.certifiedwood.org, is the place to go. CFPC is an independent, not-for-profit organization whose mission is to "conserve, protect, and restore the world's forests by promoting responsible forest products buying practices throughout North America."

CFPC's web listing of certified suppliers covers products from forests that an accredited, independent certification organization has certified as well managed, according to stringent, third-party standards. Independent, third-party forest certification is site-specific, and it validates on-the-ground operations as employing the best management practices to ensure the long-term health of the total forest ecosystem.

Logo courtesy of the Certified Forest Products Council

Domestic alternatives

Quite often you can find small mills in your area that cut up local trees into lumber. The big advantage of woods from these mills is that they're usually a lot cheaper than "standard" hardwoods. Some of the more common domestic alternatives worth checking out are ash, birch, poplar, and soft maple.

Ash for oak

I love working with ash. It's inexpensive, it smells great, and it can be stained to look just like oak. *From left to right and top to bottom in the top photo:* red oak and white oak, and ash finished with red oak stain, natural, and dark oak stain. Ash often has very straight grain, which makes it a joy to work. It's also often available in wider pieces than oak, and it generally costs 1 to 2 dollars a board foot less than oak.

Birch for maple or cherry

Depending on where you live, you may be able to find yellow birch at much lower prices than hard maple or cherry. Birch is just as hard as either of these woods and, when stained, can closely resemble cherry. (In the middle photo, the top piece is birch, the bottom is maple; the left side is natural, and the right side has one coat of satin polyurethane.) Birch is sweet-smelling, fine-grained, and easy to work.

Poplar and soft maple

Two softer hardwoods that can be used for "secondary" parts in a project like drawer sides and backs are poplar and soft maple (the two middle pieces in the bottom photo are poplar, the others are soft maple). Both species are inexpensive and work well. Poplar is particularly well suited for parts that won't show or will be painted, as it often exhibits yellow-green streaks and mineral stains.

RECLAIMED WOOD

One of my favorite alternative materials is reclaimed or recycled wood. Not only is it attractive, but as wood prices continue to rise, reclaimed wood is rapidly becoming an economical alternative. Another plus: Much of the wood "harvested" is old-growth wood—stuff you just can't get your hands on anymore, like wide heart pine and chestnut. Although there are quite a few wood recycling companies spread across the United States, they all operate in much the same manner.

Raw material

The primary sources of wood for most reclamation companies are old barns and other industrial buildings scheduled for demolition. But instead of ending up as landfill, the structures are painstakingly disassembled one piece at a time to salvage the wood, by enterprises like Sylvan Brandt in Lititz, Pennsylvania. Once taken apart, the timbers and planks are transported back to the mill's yard (*top photo*).

Checking for metal

Just like any other mill, reclamation companies will check their logs for metal before resawing (*middle photo*). The big difference here is that since the timbers and planks were used in construction, the reclaimers expect to find a lot of metal—and they do. Hammers, pry bars, and if necessary plunge cuts with chainsaws are used to remove nails, screws, mending plates, and other miscellaneous bits of metal.

NOT JUST FOR FLOORS

▨ When most folks think of reclaimed wood, they think of flooring—and this is the number one use of this wood. But reclaimed wood is also a great way to add extra character to pieces of furniture, especially country-style or Colonial. Reclaimed wood also makes for attractive trim in a home, everything from molding to wainscoting (*see photo*).

Cutting the timbers

After all the metal is removed from the beams, they're ready for the mill. Although they don't look like much in the rough and weathered stage, as soon as the first cut is made, you're faced with gorgeous wood, typically wide and clear (*top photo*). Most of the boards trimmed off the beam are cut 1" thick so that they can be planed down to ³/₄". After the boards are cut from the beam, they are moved into a simple kiln to bring each board to the same moisture content, typically around 12%, which is fine for flooring. Note: If you're planning to use reclaimed wood for furniture, have it dried down to around 8%.

Planing

The uniformly dry boards are then moved to the planer for surfacing. Some planers, like the one shown in the middle photo, will also shape a tongue on one edge and a groove on the other as the board is simultaneously being planed. After shaping, the flooring is dead-stacked on a pallet and is either shipped to a customer immediately or stored for future sale.

Finished product

The resulting product is clear, smooth flooring with a number of natural and man-made characteristics such as random nail holes (*bottom photo*). Species will vary, but at Sylvan Brandt they handle white pine, yellow pine, oak, poplar, and chestnut. Although many reclamation companies sell wholesale only, some of the smaller firms will gladly sell you a few boards, or even custom-cut some lumber for you. You can find numerous companies on the Internet by searching for reclaimed or recycled wood.

NON-WOOD PRODUCTS

There are several non-wood products that can handle some jobs better than wood: phenolic, laminate, clear plastics, and PVC.

Phenolic

Phenolic is a very strong and durable material that's made of paper and resins (*large rectangular brown plate in top photo*). It's terrific for jig making, as it's very stable and won't expand or contract with changes in humidity as wood will. This makes it especially useful as a runner to fit in the miter gauge slot of a table saw, shaper, or bandsaw, or to attach your router to for insertion into a router table. This stuff is so hard, I recommend cutting it only with a carbide-tipped blade. Wear a dust mask when cutting this: The fine dust generated not only will irritate lungs, but it smells awful, too.

Two other materials that are becoming popular for jig building are high-density polyethylene (HDPE) and ultra-high molecular weight (UHMW) plastic (*white strip and disk in top photo*). Both of these are low-friction, nonstick, and great for sliding surfaces like the fence for a router table or table saw. The big advantage of these is they're easier to work than phenolic. The only disadvantage to these opaque plastics is they can't be glued—even epoxy won't work.

Laminate

Anytime a project calls for a surface that's almost indestructible, I reach for plastic laminate, often referred to by its common brand names, Formica or Wilsonart (*middle photo*). I've used plastic laminate for years for projects for my kids: easels, travel boxes, any surface that I know will get hard use. Plastic laminate is also great for around-the-house projects like TV trays and microwave carts. But the place I use plastic laminate the most is in the shop. It's perfect for sliding surfaces like router fences and the bottoms of sliding tables. I also cover assembly

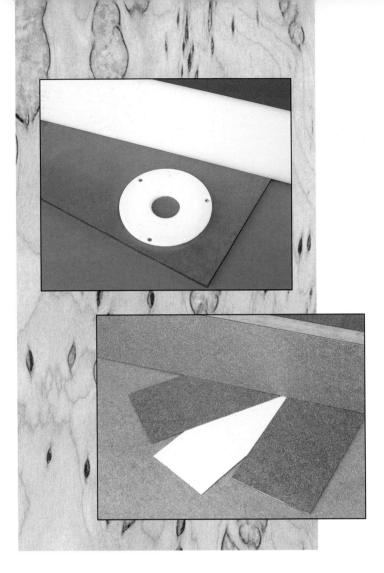

tables and glue-up fixtures with it so that excess glue can quickly and easily be scraped off.

Laminate comes in a variety of colors and is available in 4-ft. × 8-ft. and 4-ft. × 12-ft. sheets. Many lumberyards and home centers sell smaller pieces. Standard colors are stocked and others are special-ordered. Occasionally at a home center you'll find, at a discount price, a piece that someone never picked up.

Laminate cuts well with carbide-tipped tools. Just be aware that freshly cut edges are very sharp. Contact cement spread on both surfaces will create an amazingly strong bond. I recommend buying a J-type laminate roller to press the laminate firmly onto the surface. A flush-trim bit in a router will clean up protruding laminate and will leave a clean, smooth edge.

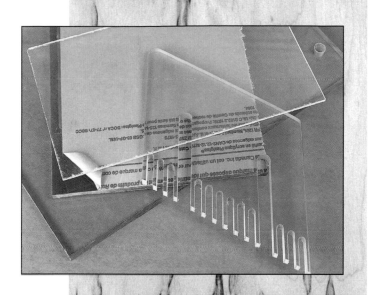

Clear plastic

I also frequently use sheets of clear plastic (polycarbonate or acrylic) when making jigs. In most cases, the purpose is some form of see-through guard, such as the blade guard on a sliding cutoff table. Thicker plastic also works well for featherboards (*top photo*). Clear plastics can be purchased from most mail-order woodworking catalogs in either clear or safety orange, in standard thicknesses of $1/8$", $1/4$", $3/8$", and $3/4$". You can cut plastic with a standard saw blade; just make sure to keep the protective sheet in place when you cut, to reduce chip-out and to prevent scratches. You can smooth rough edges with a mill file and sand them with sandpaper.

PVC

Although I most often use PVC (polyvinyl chloride) pipe for jigs, it's also useful as a base for veneering (*middle photo*). With flexible veneers, you can quickly create perfect wood columns by wrapping the veneer around a piece of PVC. If you insert toy wheels into the ends of short lengths of PVC and run a steel rod through the wheels, you can make rollers for an outfeed table for a table saw, bandsaw, or planer. PVC is also useful in the shop for storing cutoffs, dowels, and molding (*see page 129*).

OUTDOOR ALTERNATIVES

There are few woods that really hold up well outdoors—teak, redwood, white oak, and cypress among them. The next time you're building an outdoor project, whether it's a deck, a planter, or a bench, consider a solid wood alternative (*see photo*). There are a number of recycled wood/plastic composites being manufactured, like ChoiceDek and Trex, that hold up extremely well when exposed to the elements. All of these can be worked with standard woodworking tools and can be ordered from most lumberyards.

APPENDIX:
Hazards of Wood

For the most part, working with wood is a real pleasure. There are, however, some less-than-desirable side effects when working with some types of wood and wood products. Sap from some woods can cause irritation to the skin; wood and wood dust can cause hypersensitivity, allergic reactions, skin irritation, and respiratory problems (*see the chart below*).

Inhaling the dust of some hardwoods can even cause nasal (sinus) cancer. What's really deceptive about this type of cancer is that it has a latency period of 40 to 45 years. This means you could work with wood a long time with no apparent problems and then suddenly develop cancer. Over half of all known cases of this type of cancer are found in woodworkers.

Dust control

The good news is, you can prevent this from happening with proper dust control. For starters, always wear a high-quality cartridge-type dust mask when working in a dusty environment (*top photo on page* 185). Also, dust collectors will cut down on the dust in a shop; if you don't have one, you should wear your mask anytime there's dust in the air—not just when you're cutting.

COMMON WOODS KNOWN TO BE HAZARDOUS

Type of Wood	Source of Irritation	Type of Reaction	What It Effects	How potent is it? (1=low, 4=high)
Beech	leaves, bark, and dust	sensitizer	eyes, skin, and respiratory	2
Birch	wood and dust	sensitizer	respiratory	2
Cocobolo	wood and dust	irritant and sensitizer	eyes, skin, and respiratory	3
Ebony	wood and dust	irritant and sensitizer	eyes and skin	2
Iroko	wood and dust	irritant and sensitizer	eyes, skin, and respiratory	3
Purpleheart	wood and dust	(nausea)	nausea	2
Rosewoods	wood and dust	irritant and sensitizer	respiratory, eyes, and skin	4
Satinwood	wood and dust	irritant	respiratory, eyes, and skin	3
Teak	dust	sensitizer	eyes, skin, and respiratory	2
Walnut, black	wood and dust	sensitizer	eyes and skin	2
Wenge	wood and dust	sensitizer	respiratory, eyes, and skin	2
Western red cedar	leaves, bark, and dust	sensitizer	respiratory	3
Yew	dust	irritant	eyes and skin	2

Sensitizing woods

Sensitizing woods take a while to have any effect on you—basically, you become more sensitive the more frequently you're exposed. Black walnut is good example. I have a friend who used to work with walnut all the time, without a dust mask. Now if he walks into a shop where somebody is working with it, he immediately starts sneezing and his eyes water. It's a real shame because he loves the wood, and he could have prevented his reaction by wearing a mask.

Composites

The dust from composites (*left photo*) can cause respiratory problems for two reasons. First, the dust can be superfine, especially the dust generated from cutting or sanding MDF (medium-density fiberboard). Second, since all composites are bonded together with resins, the dust contains a high amount of resins. Neither the fine dust nor the resins belong in your lungs. Use dust collection, if you have it, when machining composites, and always wear a dust mask.

Exotics

Most woodworkers have some form of reaction when working with tropical hardwoods or exotics (*left photo*). I have a problem with rosewood and cocobolo; one snoutful of dust, and I can't breathe. If you're planning on working with an exotic, wear a dust mask and keep the dust away from your face, particularly your eyes. Just a little dust in your eye, and it can start watering and even swell up. Goggles are a good idea, but a full-face respirator is even better.

Pressure-treated lumber

A final wood product to watch out for is pressure-treated lumber (*bottom photo*). The dust generated from working with pressure-treated lumber contains the chemical used to treat the wood, typically chromated copper arsenate—not something you want in your body. Here again, use dust collection if you have it and wear a dust mask; whenever possible, wear gloves when handling pressure-treated lumber to keep your skin from absorbing the chemicals within.

SOURCES
and Information

Drying lumber information

Forest Products Society
608 231-1361
www.forestprod.org

Engineered products information

Composite Panel Association
301 670-0604
www.pbmdf.org

Structural Board Association
U.S.: 218 829-3055
Canada: 416 730-9090
www.osbguide.com

Grading supplies and information

Conway Cleveland
(lumber-grading sticks)
616 458-0056

National Hardwood Lumber
Association
901 377-1818
www.natlhardwood.org

Western Wood Products
Association
503 224-3930
www.wwpa.org

Mail-order wood and veneer

Constantine's
800 223-8087
www.constantines.com

Rockler Woodworking and
Hardware
(formerly The Woodworkers' Store)
800 279-4441
www.rockler.com

Woodcraft
800 225-1153
www.woodcraft.com

Woodworker's Supply, Inc.
800 645-9292

Plywood and information

APA (Engineered Wood
Association)
Western U.S.: 206 565-6600
Eastern U.S.: 770 427 9371
www.apawood.org

Hardwood Plywood & Veneer
Association
703 435-2900
www.hpva.org

States Industries
(ApplePly, prefinished plywood)
800 843-2753
www.statesind.com

Sawmill manufacturers

Granberg International
(chainsaw mill)
510 237-2099
www.granberg.com

Logosol
(chainsaw mill)
877 564-6765
www.logosolusa.com

Timberking
(bandsaw mill)
800 942-4406
www.timberking.com

Woodmizer
(bandsaw mill)
800 553-0182
www.woodmizer.com

Wood
GLOSSARY

air-dried lumber – lumber that has reached its equilibrium moisture content by being exposed to air.

annual growth ring – the visible layer of growth that a tree gains in a single year, comprised of one layer of earlywood and one layer of latewood.

bandsaw mill – a type of portable saw mill that uses a continuous band to cut the log.

bark – the outermost layer of a tree's trunk, which protects the inner wood; composed of an outer, dead cork layer and an inner, living layer.

bird's-eye – a much sought-after type of figure that's caused by indentations in the cambium layers of some wood species, most notably maple.

board foot – a unit of wood volume equal to 144 square inches; a board that's 1" thick and 12" inches square measures 1 board foot.

bookmatch – a way to match veneer where alternating pieces of veneer sliced from a log are turned over like the pages of a book and joined together.

bound water – the moisture present in wood found within the cell walls.

bow – a form of warp that is an end-to-end curve along the length of a board.

burl – a wartlike growth that forms on a tree and that, when sliced, produces extremely disoriented grain patterns that are quite attractive.

cambium – a 1-cell-thick layer of cells located between the phloem and the sapwood that is the active-growing part of a tree.

cant – a log that has been debarked and sawn square.

case-hardening – a drying defect where the surface of wood dries faster than the wetter inner core; this causes permanent set and stresses that release when the board is cut.

chainsaw mill – an accessory for a chainsaw that allows logs to be rough-cut into lumber on site.

compression strength – how well a wood holds together when sustained stress is applied.

compression wood – a type of reaction wood that forms in softwoods on the underside of a leaning trunk or limb.

conifer – a type of tree that's characterized by needle-like or scale-like foliage, usually evergreen.

crotch – the highly figured wood that occurs where a limb joins a trunk; the grain swirls dramatically where the wood fibers have crowded and twisted together.

crook – a form of warp that is an end-to-end curve along the edge of a board.

cup – a form of warp that is an edge-to-edge curve across the face of a board.

debarker – a large machine at a sawmill that removes the bark from a tree by grinding or chipping it off.

deciduous – a type of tree where the leaves fall off every autumn; typically a hardwood, but not always. Some hardwoods in tropical regions keep their leaves all year long.

diffuse-porous – hardwoods where the vessels formed throughout the growing are of uniform size and evenly distributed.

earlywood – a layer of sap-conducting wood cells that form early in the growing season.

end checks – a drying defect caused by the ends of boards drying faster than the rest of the wood; can usually be preventing by sealing the end grain.

end-coating – the process of sealing the ends of boards to prevent checking caused by unrestrained evaporation of moisture.

equilibrium moisture content – the point at which no moisture enters or leaves a piece of wood.

extractive – resins and other substances deposited in the heartwood during a tree's growth that impart both color and resistance to decay.

FAS – an acronym for "firsts and seconds," the top grade for hardwood lumber; an FAS board will yield at least $83^1/3\%$ defect-free lumber.

fiber saturation point (FSP) – a condition of wood cells where they are free from all water, but the cell walls remain fully saturated.

fiddleback – a type of washboard-like figure that occurs in some species of wood with wavy grain.

figure – the pattern on a wood's surface, resulting from the combination of its natural features and the way the log was cut.

grade – a designation of the quality of a log or wood product such as lumber, veneer, or plywood.

grade stamp – a stamp applied to lumber and other wood products that indicates the product's grade, strength properties, species, and mill where it was cut or manufactured.

grading stick – a special ruler used by hardwood-lumber graders to quickly determine the surface measure of a board so that the grade can be determined.

grain – the direction of wood fibers in a tree or piece of wood with respect to the axis of the trunk.

half-round veneer – veneer that's produced by moving a log in an abbreviated arc against a knife held roughly parallel to the center of a log.

hardboard – often referred by the trade name Masonite, it's an engineered panel made up of finely ground processed wood fibers and resins bonded together under heat and pressure.

hardwood – wood cut from broad-leaved, mostly deciduous trees that belong to the botanical group Angiospermae.

head rig – a moving carriage on a circular-saw mill that cradles a log and presents it to the blade to cut planks or timbers.

heartwood – mature wood that forms the spine of a tree.

honeycomb – a drying defect that occurs when lumber undergoes severe case-hardening in the early stages of drying; appears as deep, internal checks.

impact bending – how well a wood handles impact.

juvenile wood – the wood in every tree that forms within its first 10 years or so; usually has undesirable characteristics such as low strength and shrinkage along the grain.

kiln – a heated chamber of a building used to dry lumber; humidity and air circulation are constantly monitored and adjusted as the wood dries.

kiln-dried lumber – lumber that has dried in a kiln to a specific moisture content.

knot – the section of a branch or limb that has been overgrown by expanding girth of a tree; may be loose or tight.

latewood – a layer of small, thick-walled wood cells that form late in the growing season.

medium-density fiberboard (MDF) – an engineered panel where wood chips are further reduced to tiny fibers that are coated with resin and hot-pressed to create a homogenous panel of uniform density.

medium-density overlay (MDO) – B-grade plywood that's been covered with a smooth resin-impregnated paper overlay.

melamine – a type of particleboard with a thin layer of plastic bonded to both faces.

modulus of elasticity – the measure of a wood's ability to spring back after a load is removed.

modulus of rupture – the maximum load a wood can support without breaking.

moisture content – the amount of water in a piece of wood expressed as a percentage of the green weight minus the dry weight times 100, divided by the green weight.

moisture meter – an electronic device used to measure the moisture content of lumber.

nominal dimensions – dimensions based on rough-cut (unplaned) softwood; a 2×4 is nominally 2" × 4"—it's actually $1^1/2$" × $3^1/2$".

oriented-strand board (OSB) – an engineered panel that's made from strands of wood bonded together with a waterproof resin under heat and pressure; the oriented strands create cross-banding somewhat similar to plywood.

parenchyma – a type of specialized wood cell that serves primarily as storage for fluids and other materials.

particleboard – a wood-panel product that is produced mechanically by reducing wood to small particles, applying adhesive to the particles, and pressing a mat of particles under heat and pressure.

phenolic – a durable, strong plastic that's made of paper and resins; useful for making jigs.

pith – the small, soft core occurring in the center of a tree trunk.

pits – recesses in portions of wood cells that allow fluids to flow from cell to cell.

plain-sawn – the most common way to cut a log, where the cut is made tangential to the growth rings; also know as flat-sawn.

plain-sliced veneer – veneer produced by slicing a log with the knife parallel to the center of the log.

plastic laminate – known by the brand names Formica and Wilsonart, it is a thin, almost indestructible sheet of plastic that's most often bonded to a surface with contact cement.

pleasing match – a way of joining veneer where attention is paid to the color of the pieces but not necessarily the grain.

plywood – panels made up of thin sheets of veneer or plies glued together so the grain is perpendicular; may also have a core made of particleboard or other composite material.

quarter designation – a rough thickness designation used for hardwoods based on $1/4$" increments; a 5/4 or five-quarter board is $1^1/4$" thick before planing or surfacing.

quartersawn – lumber that has been cut so that the growth rings are between 45° and 90° to the face of the board; quartersawing some species results in ray fleck or silver grain.

quarter-sliced veneer – veneer produced by slicing a log with the knife perpendicular to the growth rings.

random match – a way of joining pieces of veneer where no attention is paid to color or grain of the pieces being joined.

ray – a ribbon-shaped strand of wood cells that extends from the inner bark to the pith perpendicular to the axis of a tree trunk; rays appear as fleck on quartersawn surfaces of some species.

reaction wood – wood that a tree forms to try to bring a leaning trunk or branch back to vertical; can be compression wood or tension wood.

reclaimed wood – lumber that has been sawn from used timbers, often harvested from old barns and commercial buildings.

relative humidity – the ratio of water vapor present in the air to the amount that the air would hold at its saturation point, usually expressed as a percentage.

resin canals – spaces between softwood cells caused by separation of adjacent cells; they serve as a defense mechanism by transporting resin to an injury.

rift-sawn – wood that has been cut so the growth rings are at an angle between 30° and 60° to the face of the board.

ring-porous – hardwoods where vessels that form early in the season are much larger than pores formed later; this forms a distinct zone of earlywood and latewood, easily seen with the naked eye.

rotary-sliced veneer – veneer produced by pressing a broad cutting knife set at a slight angle against a rotating log.

S2S or S4S – an acronym for "surfaced two sides" or "surfaced four sides"; describes which faces of rough lumber have been surfaced smooth.

sapwood – new wood surrounding the denser heartwood.

sawyer – a skilled professional who reads a log, determines the best way to cut it, and via a set of controls has the mill make the cuts.

shake – a lumber defect that is a lengthwise separation of wood, usually along the growth rings.

shear strength – a wood's ability to resist internal slipping of one part along another, along the grain.

shrinkage – changes in the dimensions of wood due to loss of moisture below the fiber saturation point.

slip-match – a way of joining veneer where successive pieces are slipped out in sequence and joined together.

softwood – wood cut from coniferous trees belonging to the botanical group Gymnospermae.

spalting – an attractive dark brown or black stain in some woods caused by decay.

specific gravity – the ratio of the density of a wood to the density of the water stored in it at a specified temperature.

speck – a defect that's caused by a fungus living in a tree, which appears as small white pits or spots.

split – a separation of wood fibers that extends completely through a piece of lumber, usually at the ends.

sticker – a piece of wood, typically 3/4" square, that's inserted at regular intervals between layers of green wood to assist the drying process.

sticker stain – sometimes called shadow, it's a stain that forms under the stickers in a stack of drying wood.

stripe – a stripe or ribbon pattern that occurs when woods with interlocked grain (which slopes in alternate directions) are quartersawn.

substrate – a piece of plywood, softwood, hardwood, or engineered panel that's used in veneering as a core or foundation.

surface checks – a drying defect that occurs when the surface dries too quickly in relation to the core.

tension wood – a type of reaction wood, found in hardwoods, that forms on the upper side of a leaning trunk or limb.

texture – the size of the cells in a wood, described as ranging from coarse to fine; often confused with grain.

twist – a form of warp where one corner of a board is not aligned with the others.

veneer – a thin layer or sheet of wood sawn, sliced, or cut from a log; paper backing may be applied to create flexible veneer.

vessels – large wood cells found only in hardwoods, which connect to form a continuous path for fluid.

wane – the presence of bark or a lack of wood from any cause along the edge or corner of a piece of lumber.

warp – any deviation of the face or edge of a board from flatness, or any edge that is not at right angles to the adjacent face or edge; the most common forms of warp are bow, cup, twist, and crook.

INDEX

METRIC EQUIVALENCY CHART

Inches to millimeters and centimeters

inches	mm	cm	inches	cm	inches	cm
1/8	3	0.3	9	22.9	30	76.2
1/4	6	0.6	10	25.4	31	78.7
3/8	10	1.0	11	27.9	32	81.3
1/2	13	1.3	12	30.5	33	83.8
5/8	16	1.6	13	33.0	34	86.4
3/4	19	1.9	14	35.6	35	88.9
7/8	22	2.2	15	38.1	36	91.4
1	25	2.5	16	40.6	37	94.0
1 1/4	32	3.2	17	43.2	38	96.5
1 1/2	38	3.8	18	45.7	39	99.1
1 3/4	44	4.4	19	48.3	40	101.6
2	51	5.1	20	50.8	41	104.1
2 1/2	64	6.4	21	53.3	42	106.7
3	76	7.6	22	55.9	43	109.2
3 1/2	89	8.9	23	58.4	44	111.8
4	102	10.2	24	61.0	45	114.3
4 1/2	114	11.4	25	63.5	46	116.8
5	127	12.7	26	66.0	47	119.4
6	152	15.2	27	68.6	48	121.9
7	178	17.8	28	71.1	49	124.5
8	203	20.3	29	73.7	50	127.0

mm = millimeters cm = centimeters